Advanced Building Envelope Components

Comparative Experiments

Advanced Building Envelope Components

Comparative Experiments

FRANCESCA STAZI, PHD
Associate Professor in Building Engineering
Department of Materials
Environmental Sciences and Urban Planning (SIMAU)
Polytechnic University of Marche,
Ancona, Italy

ELSEVIER

Butterworth-Heinemann
An imprint of Elsevier

ADVANCED BUILDING ENVELOPE COMPONENTS ISBN: 978-0-12-816921-6

Publisher: Matthew Deans
Acquisition Editor: Matthew Deans
Editorial Project Manager: Mariana Kühl Leme
Production Project Manager: Sreejith Viswanathan
Cover Designer: Alan Studholme

Butterworth-Heinemann is an imprint of Elsevier
The Boulevard, Langford Lane, Kidlington, Oxford OX5 1GB, United Kingdom
50 Hampshire Street, 5th Floor, Cambridge, MA 02139, United States

Dedicated to my beloved children
Riccardo and Alessandro

Preface

THE ENVELOPE HAS A NEW ROLE NEEDING ACCURATE EVALUATIONS

Recently, the attention is increasingly focused on envelope solutions that are able to dynamically interact with both indoor and outdoor environments and to emphasize the occupant well-being.

On the one hand, the evaluation of such markedly dynamic components is very complex involving several domains (i.e., thermal, visual, air quality) within variable boundary conditions also perturbed by users' stochastic behaviors.

On the other hand, the accurate assessment of the real in-service components behavior is becoming more and more important for the extreme performance required to the buildings.

EXPERIMENTAL BASE AND HOLISTIC-STOCHASTIC EXTENSIONS

The book focuses on different and alternative advanced solutions for the building envelope, competing with each other.

It is based on a holistic approach considering multiple envelope functions, such as indoor thermal comfort, energy saving, contribution to mitigate outdoor overheating, air quality, and mechanical performance.

The components are compared through laboratory tests and simultaneous monitoring on an on-site mock-up and on adjacent rooms inside real buildings.

The experimental data are also combined with dynamic numerical simulations so that reliable infield observations could be generalized for other uses, typologies and climates. Moreover, stochastic evaluations on users' interferences are included.

NEW ENVELOPE CHALLENGES

The comparisons regard not only traditional solutions but also nanotechnology fabrics, smart materials, and new techniques recently introduced in the market.

For example, the book compares different types of smart glass coatings (low-E vs. solar control film), alternative types of external shadings, types of windows (traditional or domotic), several types of structural adhesives for advanced curtain walls, types of nanofoams for cavity walls retrofitting, types of smart treatments for plasters, types of external claddings for highly technological facades, and so on.

AIM OF THE WORK

This study intends to deepen how the building envelope components work and to offer comparisons between alternative solutions. It identifies the optimal solutions to achieve high indoor comfort levels, energy saving and urban heat island mitigation, and to enhance the users' satisfaction. The book explains the physical principles underlying the building technologies, and draws the state-of-the-art of the latest studies, identifying possible breakthrough solutions.

Far from being a comprehensive review of different envelope technologies, the book aims to explore significant solutions including experimental data and to bridge the gap between the academic theoretical studies and the effective application of innovative solutions.

Acknowledgments

The author thanks Prof. Placido Munafò for his supervision on many research works and theses that allowed obtaining several experimental data.

The author is also grateful to Prof. Marco D'Orazio for his guidance on ventilated facades interpretation and on all the complex aspect of occupants' stochastic behavior interpretation.

Many thanks to Prof. Constanzo Di Perna for his important contribution on all the research studies.

Contents

CHAPTER 1

Transparent Envelope

1.1 BACKGROUND: TOWARD A DYNAMICALLY SELECTIVE DESIGN

The role of the transparent envelope has greatly changed. Over the past decades, the technological evolutions and the new materials have led to the design of highly glazed building facades, in which the window has moved forward on the external board of the wall and has become prevalent, until almost the disappearance of its boarder frames (Sung, 2016).

Such lightweight and transparent building envelopes have progressively spread over every geographic zone. Especially, in areas with hot summers and cold winters, this has created a number of open issues, such as occupant's discomfort for overheating, accelerated degradation of interior finishing materials, visual discomfort for glare. Moreover, the main drawbacks have concerned the energetic issue because large glazed areas have resulted in elevate energy consumption for both winter heating and summer cooling.

Today, the energy consumption in buildings to maintain indoor temperature is being recognized as the main responsible for the energy demand worldwide. The studies based on hierarchical approaches (Ma et al., 2012; Xing, Hewitt, & Griffiths, 2011) and European research work groups (Marini et al., 2014) have identified the design of energy efficient building envelopes as the first step to achieve zero carbon emission.

This has led to widespread interventions of opaque walls superinsulation, thus devolving to the windows the great part of the thermal exchanges between the outdoor and the indoor and giving to this building component a huge role.

However, bare traditional glazed systems difficultly accomplish this new task. Due to the high thermal conductance of the glass and its thinness indispensable for static requirements, the U-values of the transparent parts are sensibly higher than those of the opaque facades (Cannavale et al., 2010). An unshaded glass heat transfer could be 100 times higher than an insulated wall. Hence, a large amount of energy is dissipated through such components. For instance, in a two-stories building (30% window ratio), up to 60% is lost through the windows (Rezaei, Shannigrahi, & Ramakrishna, 2017).

On the other hand, the mere improvement of the thermal insulation properties of the glazed surfaces is far from being a definitive solution. Indeed, transparent surfaces are subject to several conflicting requirements.

In winter (and in general in cold climates), when the solar radiation is nearly always welcome and illumination level are not so high, the solar gains should be maximized, by providing high solar factor, and the radiative heat loss should be avoided, by adopting low thermal transmittance values.

Instead, during the summer (and yearly in warm climates), unwanted heat gains should be excluded to prevent the thermal overheating, thus choosing a low solar factor.

In temperate climates where both conditions are present, having hot summer and cold winters, glazing systems need to change dynamically their optical properties based on the variable environmental conditions.

Selecting an ideal level of solar radiation transmission is difficult but very important for indoor thermal comfort.

Moreover, throughout the year, the glass should provide the necessary levels of natural lighting while avoiding glare. Selecting the ideal level of visible light transmission is thus important for visual comfort. This measure may also considerably reduce the electricity consumption for artificial lighting (Rezaei et al., 2017).

In general, the maximization of one aspect between energy saving, visual comfort, and thermal comfort is deleterious for the other ones. For example, the intensification of natural illuminance will save lighting energy causing, on the other hand, glare and thermal overheating.

All these contrasting needs can be fulfilled only through the adoption of multifunctional and advanced systems, for both windows and shadings.

Advanced Building Envelope Components. https://doi.org/10.1016/B978-0-12-816921-6.00001-9

FIG. 1.1 Different types of glasses with several functions.

Highly performing glazing films and advanced coatings realized with emerging materials (Fig. 1.1) are able to respond to different environmental conditions by enhancing the glass properties, such as thermal transmittance (low-e coatings), solar factor (solar control filters), maintenance (self-cleaning films), and visual transmittance (antireflective coatings).

The innovative biomimetic approaches state that such enhanced properties could be borrowed from nature, using as source of inspiration organic shapes and the adaptive behavior of natural surfaces and organisms (Al-Obaidi et al., 2017).

For example, superhydrophobic properties of cicada wings or self-cleaning effects of the lotus plant are emulated by self-cleaning coatings; the great variability of chameleon skin is mimicked by films with spectral radiation control; the highly organized structure of nacre is adopted on some low infrared emissivity composite coating; the antireflective natural surface of moth-eye is repurposed by antireflective coatings.

Apart from thermal insulation and spectral selectivity, the current nanotechnology has pointed out new ways of addressing such problems through advanced multilayered complex nanostructures. This new challenge regards the design of smart glazing coatings capable of dynamically modifying the surface properties of the glass according to the occupants' needs and to the environmental stimuli.

These new approaches have been also adopted for sun shading devices. Hinge-less, integrated, and bio-inspired systems able to change their "phenotype" as response to various stimuli (Fiorito et al., 2016) are a new research frontier. Lightweight and elastic materials equipped with actuators can be used to configure shape-morphing elements thus considerably reducing the heat transfer and controlling the solar light.

This chapter is an attempt to provide an insight on traditional and emerging technologies and to introduce the new challenges that the most advanced works on academic field are actually facing. Moreover, the most representative solutions are discussed through comparative experimental investigations.

It is composed of five parts. Section 1.2 deals with the glazing providing the basic concepts regarding glass, multiple-glazed units, and coatings. It addresses different types of coatings based either on spectral selectivity or on the dynamic interaction with the environment or even on multiple functions optimization.

Section 1.3 stresses the importance of adopting highly insulating frames for an energy efficient window and concerns frames with different materials, deepening the pro and cons of the different types.

Section 1.4 focuses on shadings. It begins with traditional static solutions and subsequently approaches the dynamic devices by giving an insight on the latest research issues.

Section 1.5 faces the open questions and future challenges of new glazing technologies.

1.2 GLAZING

The present section introduces the glazing and coating technologies. It presents the main materials and components used for insulated glass units and subsequently discusses various types of coatings, from more conventional to highly advanced ones.

It is subdivided in four subsections, corresponding to the different current methods in developing highly energy efficient windows: structural, spectral, multifunctional, and smart design.

- *Structural- layered design* works mainly on conductive heat transfer and relies on the adoption of materials as thermal barriers characterized by low thermal conductivity and good visible transmission; this approach deals with sandwiched structures, namely insulated glazing units filled with energy efficient gas, vacuum cavities, or multilayer glazing.
- *Spectral design* instead focuses on radiative heat transfers; it generally deals with thin films, tint, or

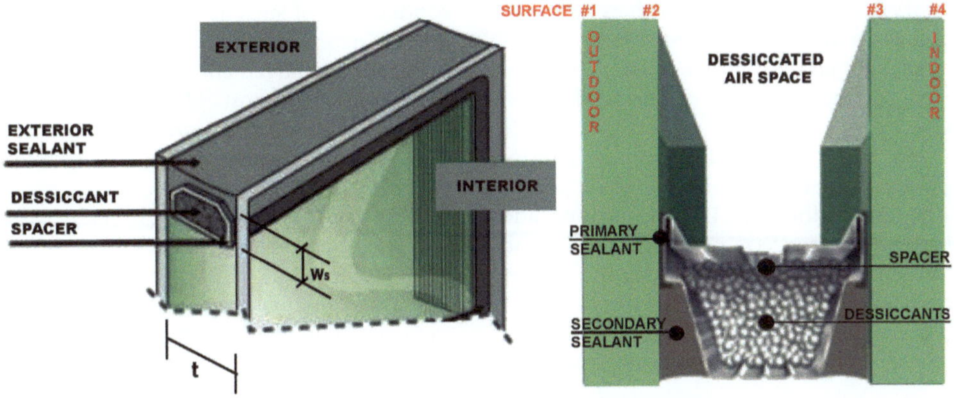

FIG. 1.2 Schemes of an insulated glass unit (IGU).

coatings able to operate a spectral selection of the incident radiation.

- *Multifunctional design* allows achieving optimal behavior on different aspects, such as maintenance, visual comfort, and energy saving as in the case of self-cleaning products.
- *Dynamic design* regards new emerging technologies that can change their properties stimulated by environmental changes or external energy sources.

1.2.1 Structural-Layered Design

This type of design includes all the solutions ranging from a simple insulated double-glazed unit IGU to multilayered glazing structures.

Insulated glass unit

Insulated glass unit (IGU), also called double-glazed unit, consists of two glass windowpanes separated by a vacuum (or gas filled) space to reduce heat transfer (Fig. 1.2). The key parameter generally used to evaluate the performance of insulated double-glazed systems is the *Thermal Transmittance U* $(W/(m^2K))$. It represents the heat loss of window due to indoor–outdoor temperature difference. U_g is the transmittance at the center of the glass (also referred to as U_{cog}); U_W instead is the overall value of the window including the frame. The detailed calculation methods are reported in Section 4.5.

The IGU comprises the following elements.

THE SPACER. A "spacer" separates the glass panes and hermetically seals the air (or gas) space between them. The typical profile width of spacer bars (w_s) varies between 4 and 8 mm. The most common spacer bar thicknesses (t) are 12 and 14 mm. Its material and position within the IGU influence the thermal performance of the window. Spacers are traditionally made of hollow aluminum (Fig. 1.3). However, metal spacers can be subject by significant heat loss and condensation because of the sharp temperature difference between the window and the surrounding air. To face such problem, spacers of a less conductive material such as thermoplastic and structural foam are proposed by the latest research studies. Data Sheet 1.A compares experimentally a warm edge spacer (realized with plastic and steel alloy) and a silicon-based superspacer.

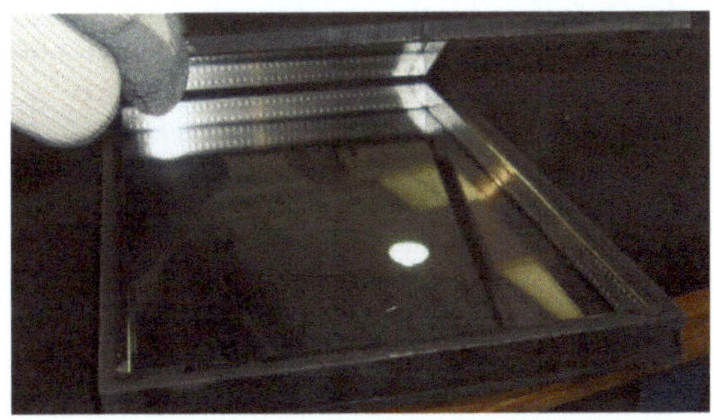

FIG. 1.3 An open IGU to make the internal aluminum spacer visible.

DESICCANT MATERIAL. Holes are drilled in the spacer bar, which is filled with a desiccant material such as silica gel or zeolite to remove moisture trapped in the gas space during manufacturing, thereby lowering the dew point of the gas and preventing condensation at temperature falling. Commonly used desiccants in the IGU industry are molecular sieves or a blend of silica gel with molecular sieves. In recent research, nanotechnology fillings such as silica aerogel are under study. In Data Sheet 1.A the traditional desiccant is experimentally compared with aerogel.

PRIMARY SEAL. A primary sealant such as rubber or butyl is applied to the sides of the spacer bar. It minimizes moisture permeation, increases UV resistance, and reduces the potential for chemical fog in the airspace.

SECONDARY SEAL. A secondary sealant such as silicon is applied along the sides of the whole unit. It provides long-term adhesion, UV protection, and greater resistance to the effects of solvents, oils, and short-term water immersion. Silicone is recognized as the best sealant for resisting weathering and adhering to glass substrates. Because of its structural properties, silicone provides structural integrity of the IG units.

IGU SURFACES. The IGU consists of four surfaces. Surface 1 is that side of the glass lite that faces the exterior of the building. Surface 2 is the other side of Surface 1. Surface 4 is the side of the glass lite that faces the interior of the building. Surface 3 is the other side of Surface 4. Each adjunctive layer brings surfaces with increasing numbers on the inside (Fig. 1.4).

GLASS. Soda—lime glass, also called soda—lime—silica glass, is the most prevalent type of glass, used for windowpanes. It is an inorganic material composed of about 70% silica (silicon dioxide), 15% soda (sodium oxide), and 9% lime (calcium oxide), with much smaller amounts of various other compounds. The glass thickness is in range of 3—10 mm or more in special applications. Laminated or tempered glass may also be used. Most IGUs present the same thickness of glass on both panes but wide ranges of thicknesses can be adopted for special (i.e., acoustic) applications.

GAS FILLING. The traditional and cheapest gas is air. Filling the window cavity with low conductive gases such as argon, krypton, or xenon instead of air is beneficial to reduce heat transfer by conduction and convection. Krypton or Xenon-filled windows produce lower U values with smaller cavities for the notably low thermal conductivity of such gases, but determine higher costs and the necessity to resolve sealant durability issues toward the improvement of the gas retention performance. Current cost of argon-filled double-glazed windows in Europe is in the range of €300—500 depending on frame configuration and labor cost (Cuce, 2018).

GAP WIDTH. U-value decreases as gap width increases. However, the decrease in the U-value beyond gap width of 12 mm is insignificant (Arici & Kan, 2015). Moreover, the optimum distance between each glass pane to minimize heat transfer depends on the filling gas. In wider cavities, air is warmed by the inner pane and rises for buoyancy forces. This creates a convection current that transfers heat from hot to cold pane. Such heat transfer mechanism could be neglected for cavities smaller than 20 mm.

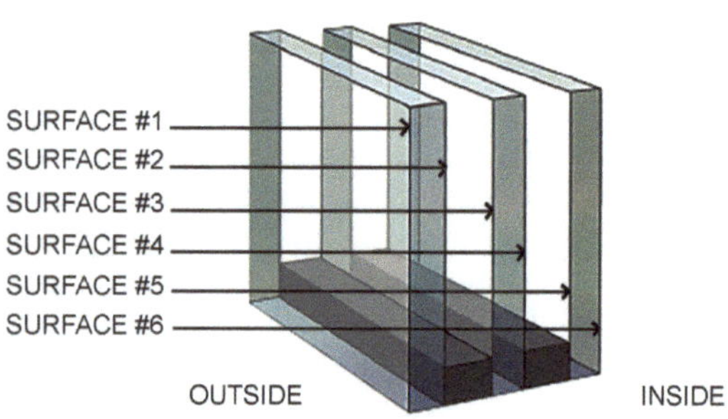

FIG. 1.4 Surface numbering in an IGU.

Multilayer glazing

The number of glazing and the inert gas type greatly affect the thermal performance of the glazing, lowering the U value.

A triple glazing product krypton filled (4 mm glass + 12 mm krypton +4 mm glass + 12 mm krypton + 4 mm glass) could reach U_g values down to 0.49 W/m^2K (Cuce & Riffat, 2015).

A lower value, down to 0.4 W/m^2K, can be reached by installing quadruple pane windows with low emissivity coatings and filling the cavity with argon (Arici & Kan, 2015).

Fig. 1.5 shows U_{cog} values (named Usi in the figure) for different fillings, number of panes, and presence of coatings/foils. It can be noted that every additional coating or pane added to the glazing unit has a diminishing effect in the U_{cog} value.

The cavity thicknesses between 8 and 16 mm were found to be optimal for IGUs with four or more panes. Only small variations within the 8—16 mm range were found. Hence, cavities can be kept at 8 mm in multilayer IGUs to maintain the total thickness of the IGU as thin as possible.

Suspended films and vacuum glazing

One disadvantage of IGUs with several glazing layers is the increased weight. A solution could be the use of nonstructural thinner intermediate layers in the IGU, maintaining the outer panes as thick as needed to ensure structural integrity. Using suspended foils (for example thin polymer films) is a promising window technology that is starting to permeate the market (Heiselberg, 2016). Using them within a double-glazed unit, some manufacturer reached 0.68 W/m^2 K

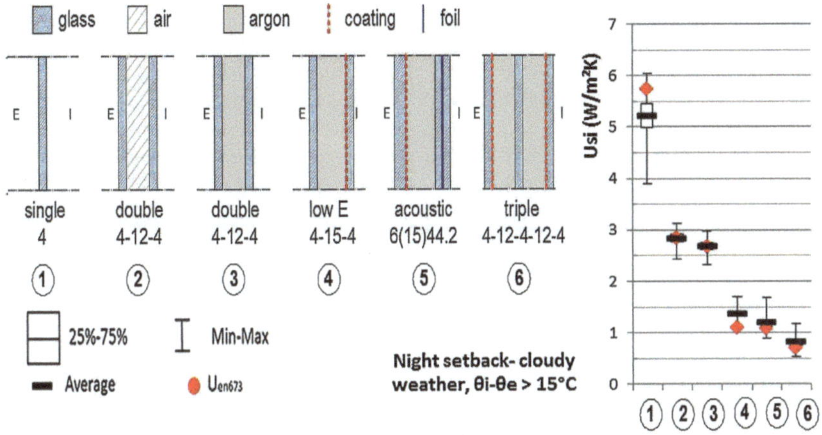

FIG. 1.5 Variation of the thermal transmittance measured on the internal surface (corresponding to Ucog) with respect to the number of glass panes and with the insertion of coatings/foil. (From Maroy et al. (2017).)

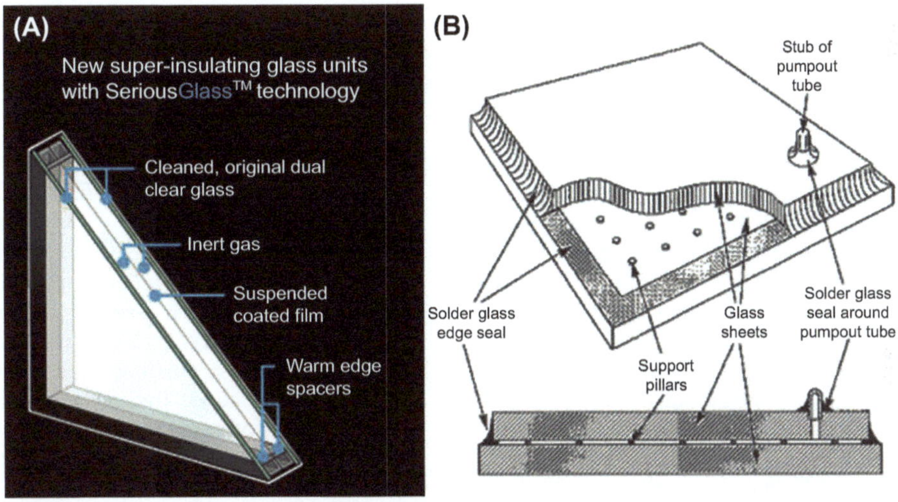

FIG. 1.6 **(A)** The Empire State Building existing windows were removed, cleaned, and retrofitted with an intermediate low-e film; **(B)** schematic diagram of a vacuum glazing. (From Jelle et al. (2012).)

(Serious Windows' 925H, North America, Fig. 1.6A). Such films are specially treated to resist ultraviolet degradation, are heat shrunk, and remain flat.

To go below the heat transfer values above, it is necessary to diminish the nonradiative (conductive and convective) heat transfer. This is the main function of evacuated "vacuum glazing" able to reach U_g values down to 0.2 W/m^{2}K in triple glass configuration (Manz, Brunner, & Wullschleger, 2006) with a reduced thickness (about half with respect to an ordinary multilayer glass with the same U value) and a light weight.

The component consists in two glass panes divided by a narrow cavity that hosts (invisible) supporting pillars (Fig. 1.6B). It can be advantageously used on retrofit interventions.

Double skins

Double skin facades (DSFs) are based on a multilayer principle. They consist of an external glazing, an intermediate gap, and an inner façade. The outer layer, generally a toughened or laminated safety glass, provides protection against the weather and hosts ventilation openings.

The air within the gap is subject to an ascending flow activated by solar-induced thermal buoyancy and wind pressure. The cavity depth varies from 10 cm to more than 2 m. The innermost glazing is commonly a double glazing that provides protection against thermal losses and is often openable to permit natural ventilation. Shading devices are usually hosted within the cavity with positive maintenance cost reduction.

The main advantage of this technique is that the air gap behaves as a thermal cushion in both summer and winter thus resulting in significant energy saving. The drawbacks lie on the fact that when the inner skin of the façade is fully glazed, the cooling loads will increase in proportion of the larger area of glazing. Moreover, the hermetic sealing could lead to condensation phenomena on the inside face of the outer skin if air from the internal space enters an inadequately ventilated gap.

A rather recent review on DSF was aimed at identifying the most influencing parameters on their thermal behavior (Barbosa & Ip, 2014). Apart from the site parameters, the study mentioned the cavity geometry, opening dimensions, the presence of massive materials in the inner skin and the type of glazing adopted. The cavity should be narrow and as tall as possible to enhance the stack effect. The application of an outer glazing with high transmittance (single glazing instead of double glazing) tends to pronounce the airflow in the cavity by increasing its air temperature. On the other hand, double glazing with high thermal insulation is likely to be applied in the inner side to reduce radiative and conductive heat transfer. The adoption of a thermally improved glazing (i.e., low-e) could even more improve the final behavior. In general, a suitable combination of glazing types is necessary to achieve an optimal balance between the enhancement of stack effect in the cavity and the reduction of heat transfer toward the inside.

The Data Sheet 1.B reports experimental measures on the comparison between a simple IGU and a double skin.

Aerogel-filled cavities

Silica aerogel is a promising material for applications in building envelopes because of its light transmitting nature and low thermal conductivity.

Aerogel was primarily adopted in Aerospace and Automotive industry. It is an excellent insulator because it is composed of gas, which is a poor conductor.

Silica aerogel consists of cross-linked SiO_2 particle chains with a large number of air-filled pores. It is derived from a gel in which the liquid part is extracted by supercritical drying and replaced by a gas.

It can be found in translucent granulates and monolithic panels. The latter generally show superior thermal and optical performance but are more expensive and require long producing times. Granular aerogels are cheaper and easier to produce, so they dominate the market despite the inferior properties.

Data Sheet 1.C surveys the convenience of their use within an insulated double-glazed unit in granular form. In particular, applications with the complete filling of the air cavity and with the use of an aerogel-filled curtain inserted within the cavity are compared.

1.A DATA SHEET: WARM EDGE OR SUPERSPACER?

The present data sheet compares alternative spacers for IGU units, all characterized by optimal performance (Fig. 1.7). In particular, it considers the following steps:

- A warm edge spacer (WES) made with a combination of highly insulating plastic and high-grade stainless steel alloy. The rectangular hollow chamber is filled for comparison either with a desiccant (sample WES des) or with granular aerogel (sample WES Aer).
- A flexible superspacer in extruded silicone foam with incorporated desiccant and preapplied side adhesive ($\lambda = 0.16$ W/(mK)).

FIG. 1.7 IGU for laboratory tests: **(A)** filling with desiccant; **(B)** sample insertion in an insulated box for the realization of infrared thermal images; **(C)** schemes of tested spacers.

Both are born to minimize the thermal bridge at the glass edge ("warm edge") and reduce the consequent risk of condensation.

All the IGU samples (3 for each type) tested were 4 mm/15 mm/4 mm, had a square shape of dimensions 40 × 40 cm, and were argon filled (90%) with a low-e glass on surface#3 and a clear glass.

The desiccant used here (32 g for each IGU side) is a porous, crystalline aluminosilicate in beaded form.

The aerogel (2.6 g for each side) was those granular produced by Cabot Corporation (dimensions from 0.5 to 4.0 mm and $\lambda = 0.018$ W/(m K)).

Although the superspacer and warm edge spacer are well known in the literature, the adoption of granular aerogel was experimentally explored until today only within the IGU air cavity instead of within the spacer bar. It is a novel application with good potential given the high thermal performance of aerogel and its drying ability. Extended results can be found in (Fiorelli, 2013).

Laboratory measurements of the thermal transmittances with imposed ΔT of 15°C (20°C on hot and 5°C on cold plate) allowed obtaining the values (referred to center of glass "cog" and sample edge) in Table 1.1. The surface thermal resistance of glass panes is set as 0.04 and 0.13 m^2K/W for the cold and warm side, respectively.

The Ucog (at the center of glass) value is significantly improved when the metal spacers are replaced with nonmetal foam based spacers. Super spacer resulted the best solution in terms of thermal performance, as it reached the lowest thermal transmittance value with reasonable costs. The aerogel filling resulted too much expensive for the reached performance.

The average values of glass surface temperature on both IGU sides in correspondence to the spacers and at midheight (Fig. 1.8) revealed the superior properties of the superspacer with very higher values especially at the edge, confirming the findings of another author (Elmahdy, 2006).

TABLE 1.1
Thermal Gradient, Thermal Transmittance, Linear Thermal Transmittance, and Costs of Warm Edge Spacer (WES) and Superspacer (SuperSp).

	ΔT cog; ΔT edge (°C)	U cog (W/(m^2K))	ψ (W/(mK))	Costs (€/m^2)
WES des	8.8; 5.9	1.38	0.024	2
WES Aer	9.1; 5.5	1.23	0.020	37
SuperSp	9.2; 2.8	1.16	0.017	9

1.B DATA SHEET: IGU OR DOUBLE SKIN?

The present data sheet regards an experimental study on a double skin facade (Fig. 1.9), compared to an unshaded insulated glass unit IGU 4/9/4 mm. The on-field monitoring lasted about 3 weeks (at the end of the summer, from September 17 until September 29, and at the beginning of the winter, from October 24

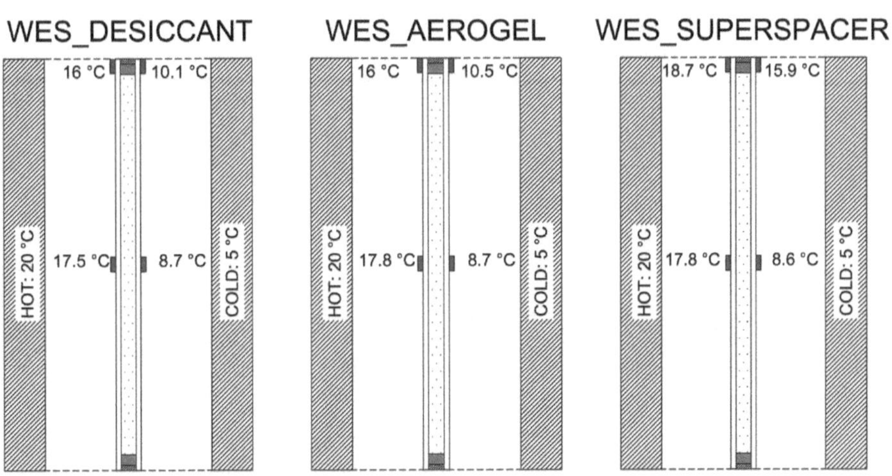

FIG. 1.8 Average temperatures measured on both sides of the three tested IGUs.

until November 6. The two glazing systems were adjacent on the building south-west side and a low emissivity coating was adopted in both cases (on the outer facing side of the inner pane). The air gap of the DSF has a width of 0.20 m. The monitoring plan is detailed in Section 4.3.1, and other results can be found in Stazi et al. (2014).

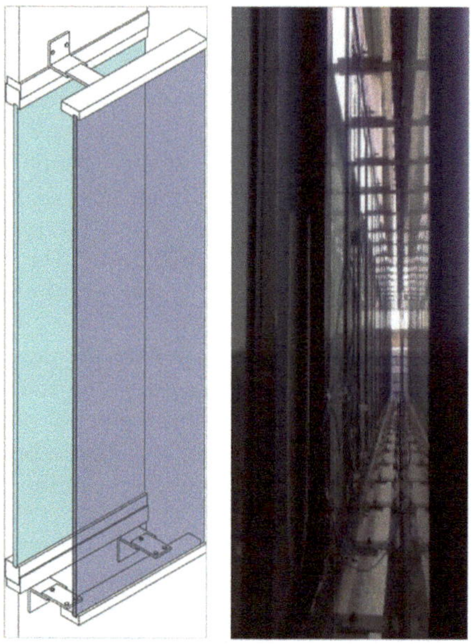

FIG. 1.9 Scheme and image of the experimented double-glazed skin.

Fig. 1.10 shows the temperatures recorded at the end of the summer on the external surface of the insulated glass unit (IGU) and on both the surfaces of the double-glazed (DG) skin facing on the channel, as well as the air temperature value within the channel of the latter one. The IGU recorded the highest values during the whole day. The ventilation within the gap lowered the temperature in an effective way (down to −10°C on the peaks).

Fig. 1.11 shows the same temperatures recorded at the beginning of the winter. On the outer glass surface, the IGU recorded higher values than the DG skin throughout the day. However, the internal glass of the double glazed skin recorded values similar to IGU except for the central day hours, when the overheating peaks are beneficially reduced.

1.C DATA SHEET: AEROGEL GRANULATES OR CURTAINS?

The present data sheet compares experimentally IGU samples in which the cavity either is filled in different manners (air, argon, aerogel granulates) or hosts a thin aerogel-filled curtain. The data are more extensively reported in (Illuminati, 2013).

Lumira^{Tm} Translucent Aerogel granulates produced by Cabot Corporation were adopted (Fig. 1.12). The studied granules have dimensions ranging from 0.5 to 4.0 mm; the declared thermal conductivity at ambient conditions was 0.018 W/(mK). Pores diameter was

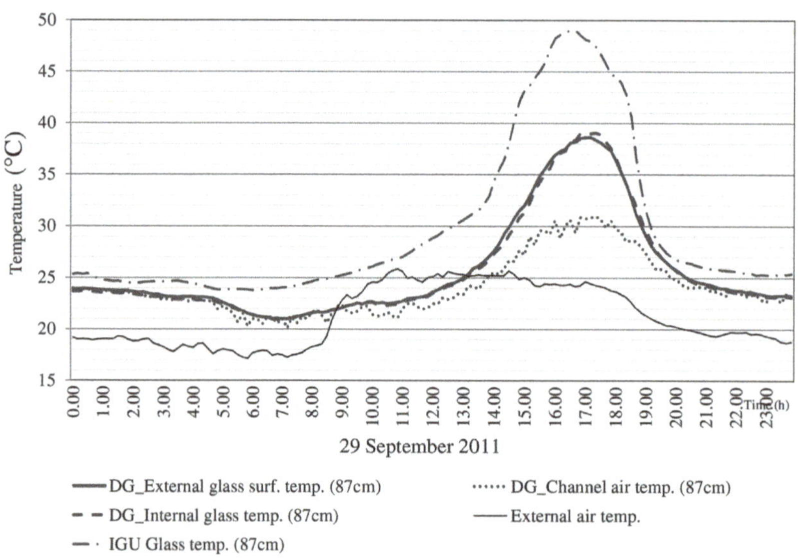

FIG. 1.10 Summer performance of the double glazed (DG) skin: experimental comparison with a traditional insulated glass unit (IGU).

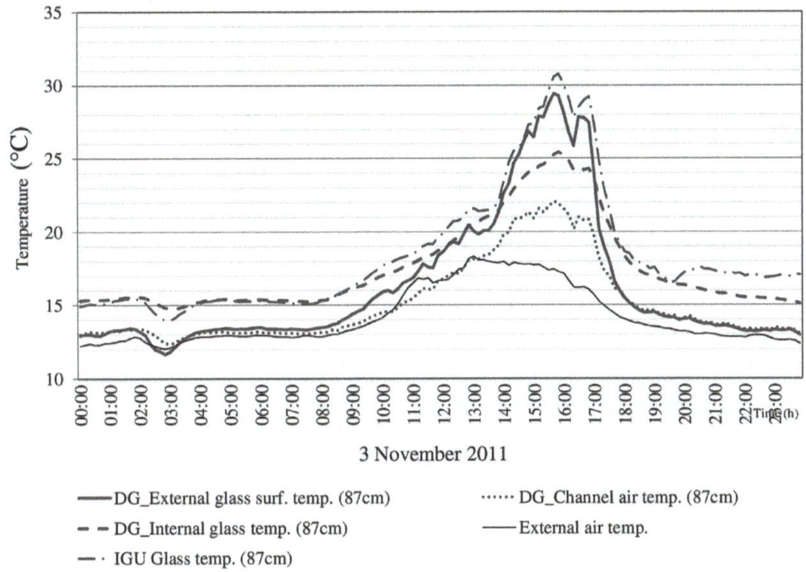

FIG. 1.11 Winter performance of the double glazed (DG) skin: experimental comparison with a traditional insulated glass unit (IGU).

20 nm and are filled with air. Bulk density ranged between 60 and 80 kg/m^3.

Ten IGU samples were realized (Fig. 1.13). All of them were 4/15/4 mm square IGUs of 40 cm × 40 cm, mounting a low-e glass ($\varepsilon = 0.03$) on surface#3. The adopted spacer was that experimented in Data Sheet 1.A, named "WES_des".

The aerogel samples were realized by opening the IGU, filling the cavity with granulates and sealing it. The procedure was realized in the laboratories of an IGU manufacturer to apply the same sealing butyl as the other samples (Fig. 1.14).

The aerogel curtain was realized by filling with the same granulate an existing polyester curtain with a honeycombed horizontal section, commonly used for applications inside the cavities (Fig. 1.15).

The thermal flux meters method was adopted for the study (Fig. 1.16). The method (see Section 4.2.1) provides the measuring of the crossing thermal flow in steady-state conditions under an imposed thermal gradient of 20°C. External and internal surface temperatures were measured on both center of glass and edge of glass area.

The results (Fig. 1.17) indicate that the solution with aerogel granulates in the internal cavity achieves the lowest thermal transmittance value. It has the highest temperature in the center of glass so the greatest thermal gradient (ΔT) with respect to the edge of glass temperature. This temperature distribution was highlighted also by infrared images.

FIG. 1.12 Aerogel granulates inserted within the IGUs cavities.

The lowest linear thermal transmittance was reached by argon-filled sample.

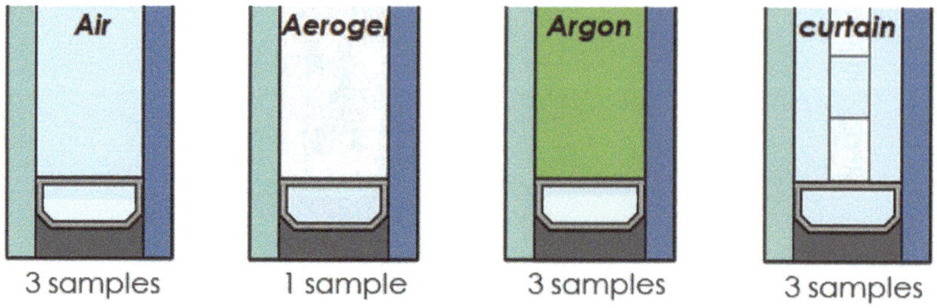

FIG. 1.13 Number and types of samples realized for the experimental comparison.

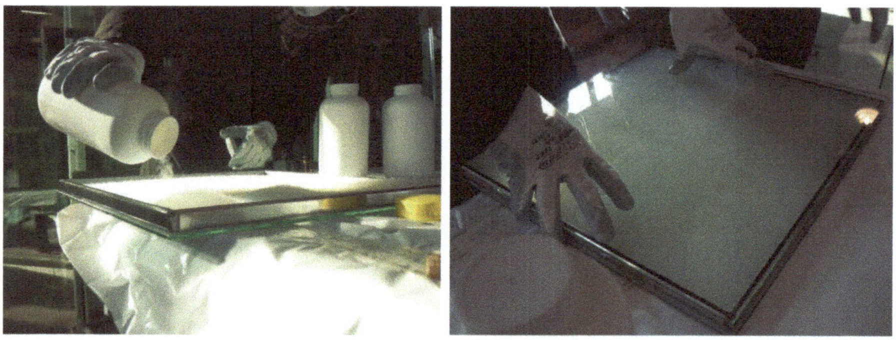

FIG. 1.14 Laboratory phases for the insertion of aerogel within the cavity.

FIG. 1.15 Different phases for the insertion of aerogel curtains within IGUs.

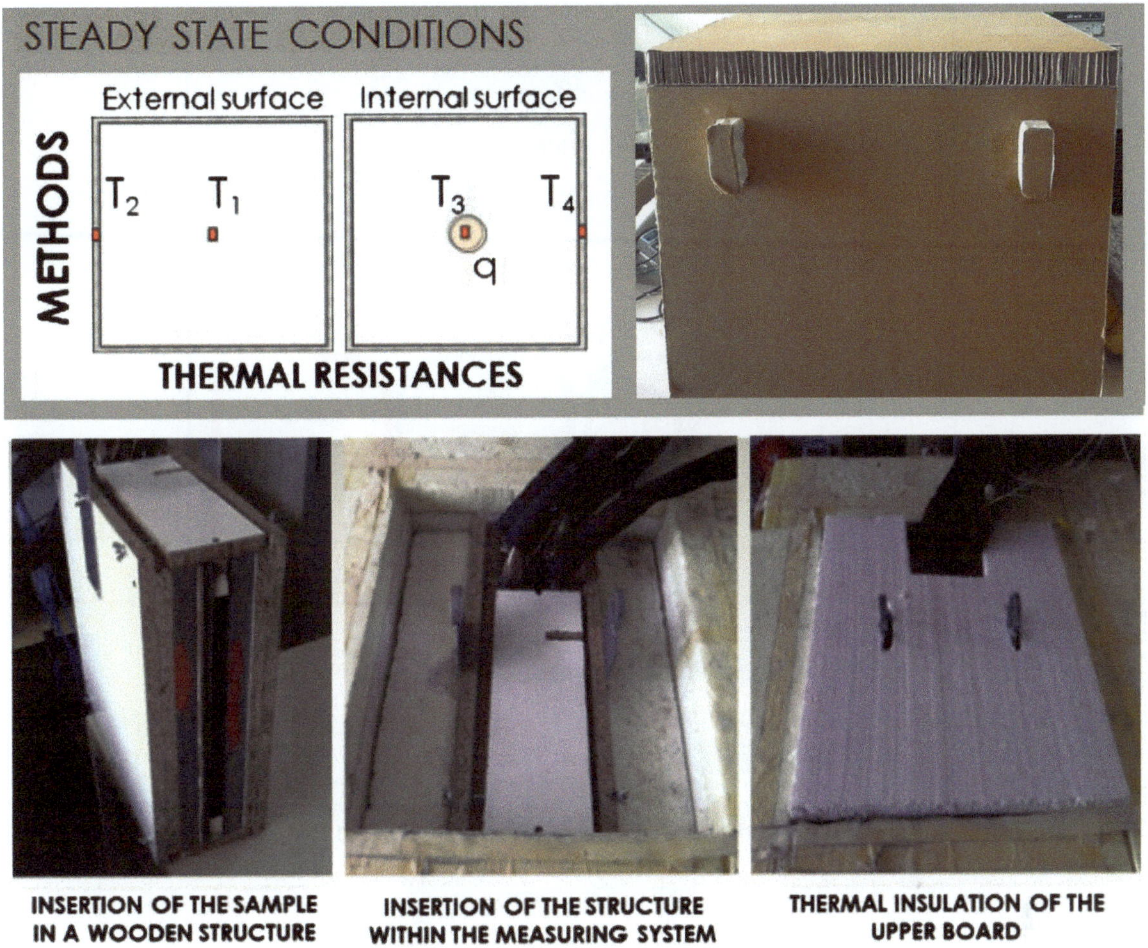

FIG. 1.16 Tools for measuring the thermal flow in steady state conditions.

Fig. 1.18 shows the temperature distribution along each sample obtained through thermography.

The graph demonstrates the best thermal performance of both aerogel-based samples.

The granulate silica aerogel application results in about 30% reduction in center of glass thermal transmittance and increased temperatures of about 2°C.

This result is consistent with the findings of other studies focused on silica aerogel for conventional glazing applications (Buratti & Moretti, 2012). They have shown that such product could achieve a 25% reduction in heat loss, outstanding acoustic insulation properties but also a 66% reduction in light transmission.

Therefore, despite the great potential of such material, the very low light transmittance of such type of solution suggests its adoption only in selected applications. This limitation and the elevate prices are the main causes of the scarcity of commercially available products.

1.2.2 Spectral Design

To understand the glazing technologies, it is important to analyze the solar energy spectrum.

Fig. 1.19 introduces the electromagnetic radiation of our ambience in a unified manner, including on the left side the solar spectrum for terrestrial solar radiation (Wang & Shi, 2017). Sunray consists of visible and invisible radiation, and it is distributed over a broad spectrum of wavelengths.

For temperatures of practical interest in the building energy studies, thermal radiation fluxes occurring in building related phenomena are within four basic bands (Table 1.2):

- Ultraviolet light, UV. It is the cause of the main degradation phenomena of indoor materials; has wavelengths λ comprised between 100 and 400 nm
- Visible light. It occupies wavelengths from about 400 to 700 nm.

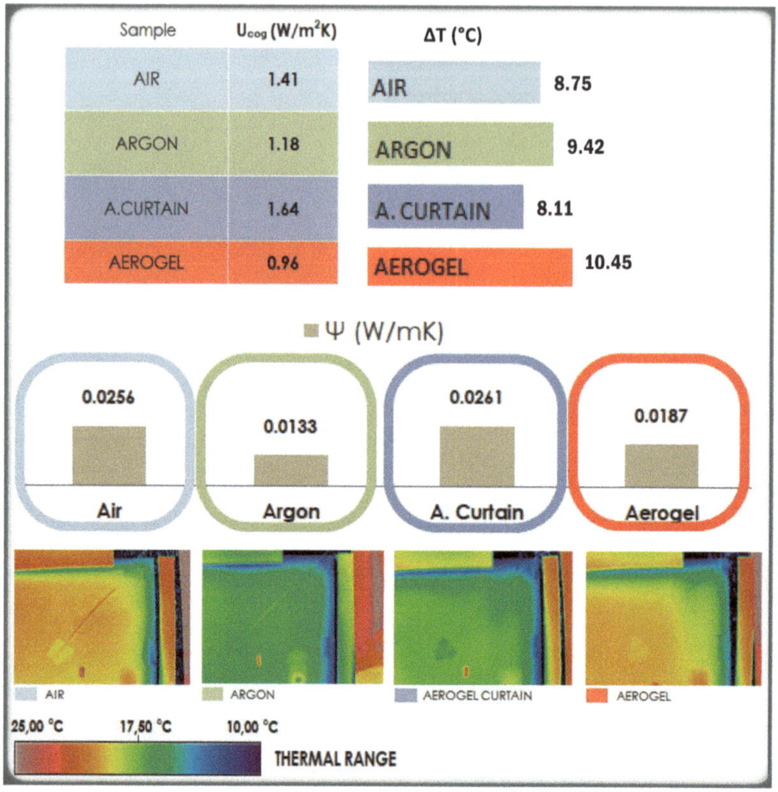

FIG. 1.17 Experimental comparison between the samples: thermal transmittances; thermal gradients ΔT between center of glass and edge of glass; linear thermal transmittances ψ and thermal infrared images.

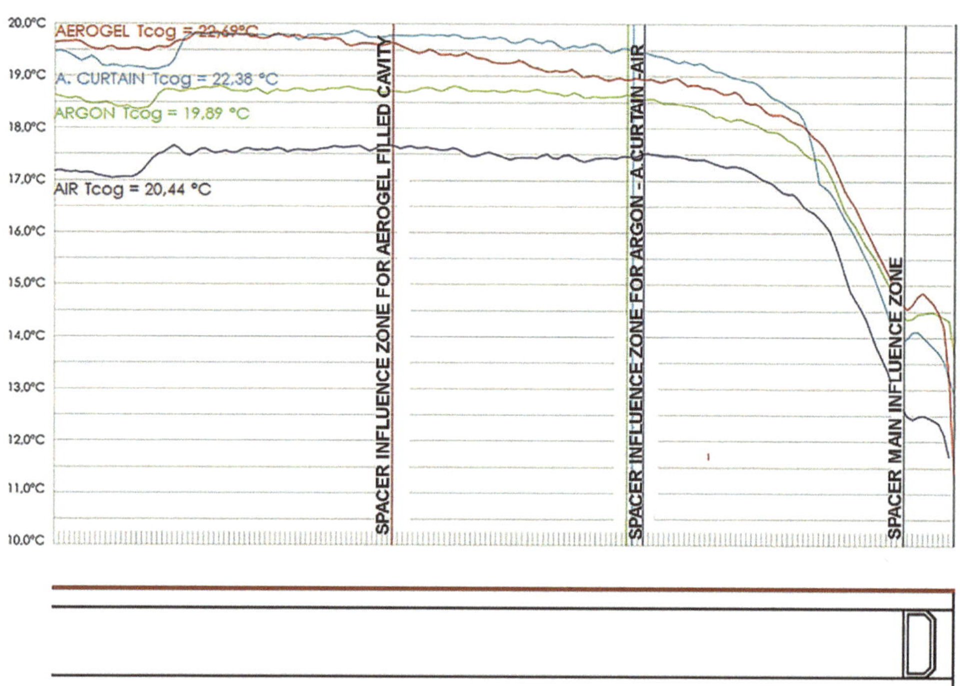

FIG. 1.18 Temperature distribution along the tested samples.

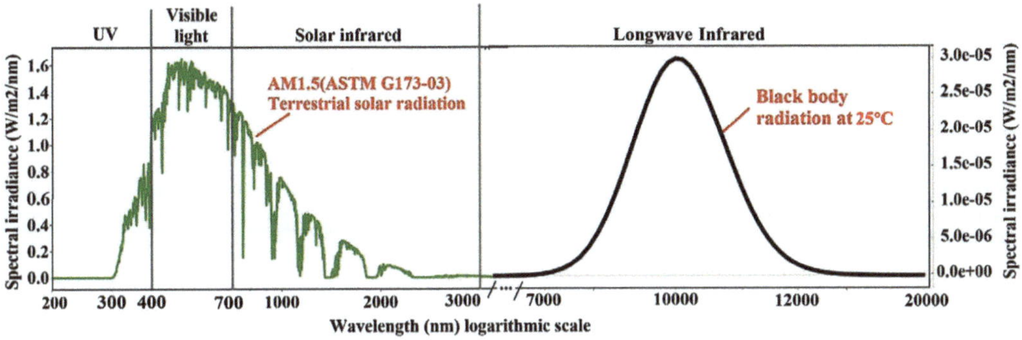

FIG. 1.19 Standard terrestrial reference spectrum and black body spectra at 25°C (it is calculated from Planck's Law). (From Wang and Shi (2017).)

TABLE 1.2
Thermal Radiation: Fluxes of Interest for Building Applications.

Spectral Region	Controlled Parameter
Ultraviolet	Materials durability
Visible	Comfort, glare, privacy
Solar infrared	Solar heat gain
Longwave infrared	Warming radiant heat energy

- Infrared light (or heat energy). It is transmitted indoor as heat; begins at wavelengths of about 700 nm. It can be subdivided in
  - Solar infrared radiation (700 nm < λ < 3000 nm), commonly referred to as shortwave infrared energy from the sun
  - Longwave infrared radiation (3.000 nm < λ < 50.000 nm). It has higher wavelengths than the sun; it is the heat emitted by terrestrial warm objects behaving to both interior and outdoor environment.

The latter, longwave radiation, is thus the thermal radiation emitted from all the materials. The spectrum reported in the figure is those obtained by a black body at 25°C.

The black body is an ideal object that absorbs all the radiation incident on it, and it is an ideal emitter showing the maximum possible emission for a given temperature. The perfect black body does not exist even if many natural substances behave nearly like it.

This ideal black body emits radiation whose spectrum—known as the Planck spectrum— is uniquely defined at a certain temperature (Fig. 1.20).

This spectrum is bell-shaped and confined to the 2<λ < 100 μm range. Thermal radiation from each material is obtained by multiplying the Planck spectrum by a numerical factor—the emittance— that is less than unity. The peak in the spectrum moves toward shorter wavelength as the temperature goes up (Wein Displacement Law) and lies at about 10 μm for room temperatures. This is because the higher the temperature the faster the electrons vibrate and the shorter will be then the emitted radiation wavelength. The black body has an emissivity of 1 while all real objects have emissivity less than 1.

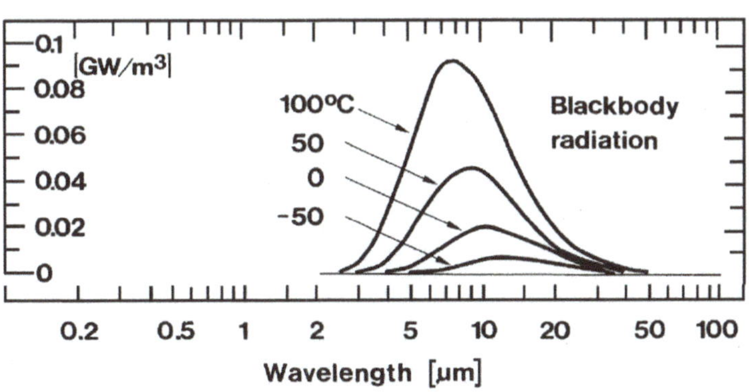

FIG. 1.20 Spectra for black body radiation pertaining to four temperatures. (From Granqvist (2007).)

It is important to note that as all the materials spectra will lay below the spectrum of black body radiation, there will be no overlap between solar radiation spectrum and the longwave thermal radiation spectrum.

Hence, it is possible tailoring surfaces with different response with regard to thermal and solar radiation. Solar energy from invisible solar spectrum comprises more than 50% of solar energy, so controlling this range is fundamental for energy saving.

When a sunray hits the glass, it is either transmitted, reflected, or absorbed depending on glass optical properties for the various wavelengths. Regulating window transmission, absorption, and reflection properties to control the passage of thermal radiation in these ranges can significantly improve building energy efficiency.

The parameters representing the optical properties of the glazing can be categorized into two families, solar light factors and solar energy factors. Their detailed explanation is reported in section 4.6.

A conventional clear glazing has high transmittance within the solar spectrum: a significant amount of solar radiation passes through the window, and heat from objects within the space is reradiated back into the glass, then from the glass to the outside of the window. By applying thin coatings to the glazing, the spectral behavior of the glazing can be changed to fit the different applications. High performance films are spectrally tailored medium that blocks a selected part of radiation.

Perfect coatings

In COLD climates that require heating all year around or in winter season of the temperate ones, an energy efficient window should transmit as much of the incoming visible and solar infrared radiations as possible and reflect the longwave infrared radiation back to the interior for energy saving. Hence, the window will "collect" as much energy as possible from the sun and loose as little energy as possible to the outside. A glazing design for maximizing energy efficiency would ideally allow the entire solar spectrum to pass through, but would block the reradiating of heat from the inside of the space (Fig. 1.21).

The ideal windows in such climates (Fig. 1.23, line 2) should have a high transmittance within the solar regions (high T_e) and as low emissivity as possible thus exhibiting high reflectance in the longwave thermal spectral region. This type of window is commonly referred to as a *low-emissivity window*.

In WARM climates, with recurrent overheating problems, the part of the solar radiation that is outside the visible spectral region can be blocked thus impeding unwanted solar heat gains (Fig. 1.22).

The ideal windows in such climates (Fig. 1.23 line 1) should have high transmittance within the visible spectral region (high T_v) and low elsewhere. About half of the solar energy is confined to the visible region, which means that about half of the solar energy can be reflected from the window without affecting its visual performance. This type of window is commonly referred to as a *solar-control window*.

Both coatings have low emissivity properties (in the longwave region) while the main difference between them is the behavior on the solar infrared range, described by the solar heat gain coefficient (g), which should be conveniently high for cold climates and low for warm ones (see the description of g in section 4.6).

In TEMPERATE climates where both conditions are present, having hot summer and cold winters, the glazing systems need to be positioned in an intermediate condition or to change dynamically their optical properties based on the variable environmental conditions.

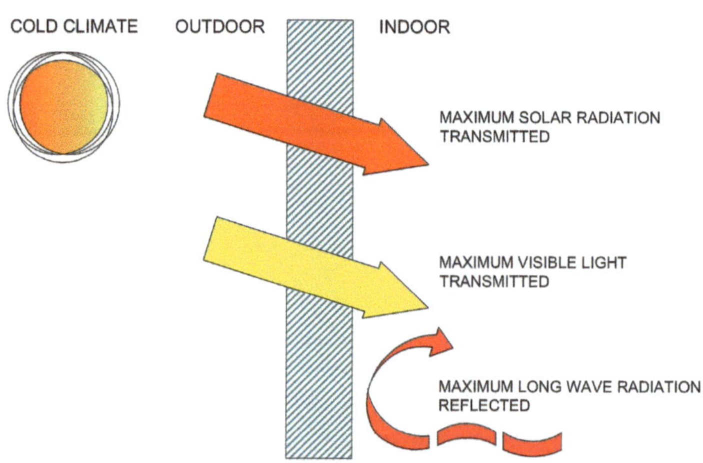

FIG. 1.21 Features of an ideal window for the winter period and in cold climates.

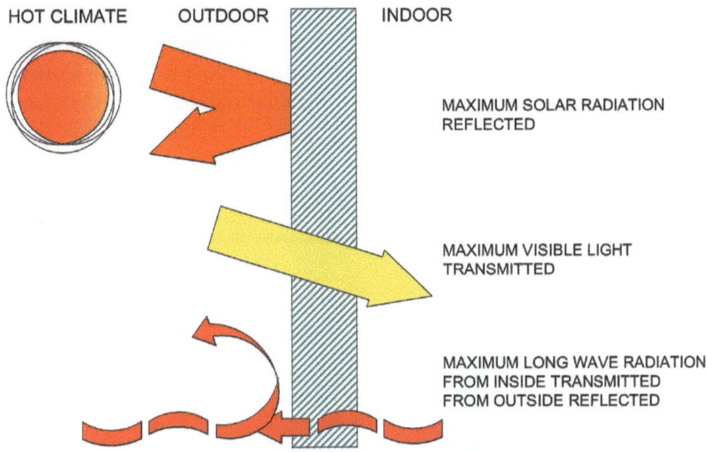

FIG. 1.22 Features of an ideal window for the summer period and in hot climates.

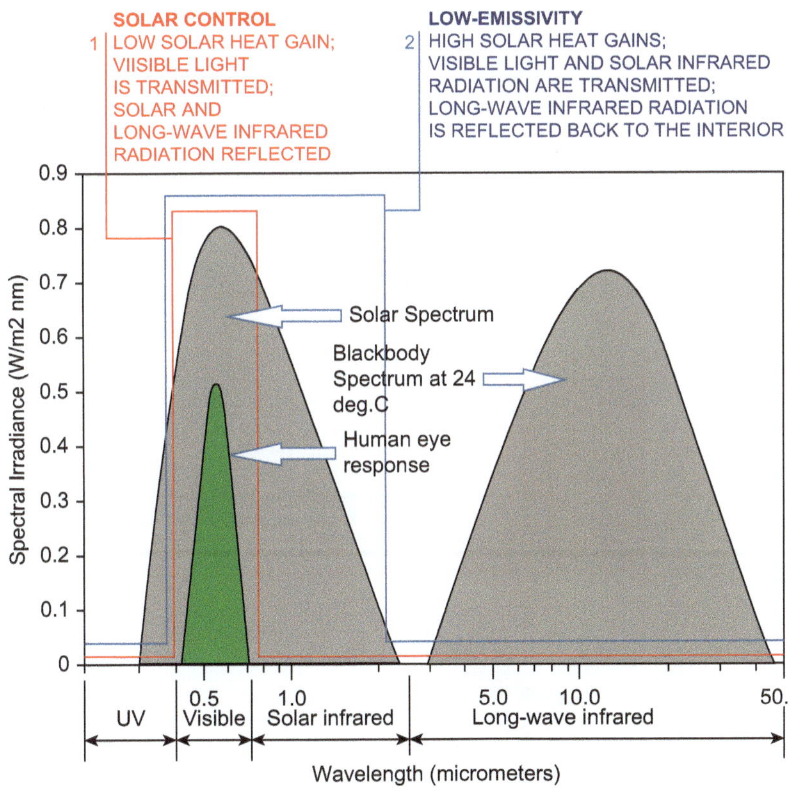

FIG. 1.23 Ideal window. (From (Kamalisarvestani et al., 2013). The scheme was initially proposed by McCluney (R, Center FSE). Fenestration solar gain analysis. Citeseer (1996).)

In every climate, the windows should have two main features: as low as possible thermal transmittance and as high as possible visible light transmission ($T_v > 0.7$). The former aspect is very important for cold climates while for the latter a lower performance can be considered satisfactory in hot conditions ($T_v > 0.6$).

Real coatings

The real spectra slightly differ from those idealized for the perfect window especially in near infrared (NIR) region (Fig. 1.24). The difference is more evident in low emissivity films that difficulty can achieve a perfect transition from high solar infrared transmittance and low transmittance in longwave region. This is a challenge that the most advanced researches are addressing by working on destructive optical interference and photo thermal nanostructured technologies.

Fig. 1.25 relates the visible transmittance values T_v and solar factors g of seven types of high performance

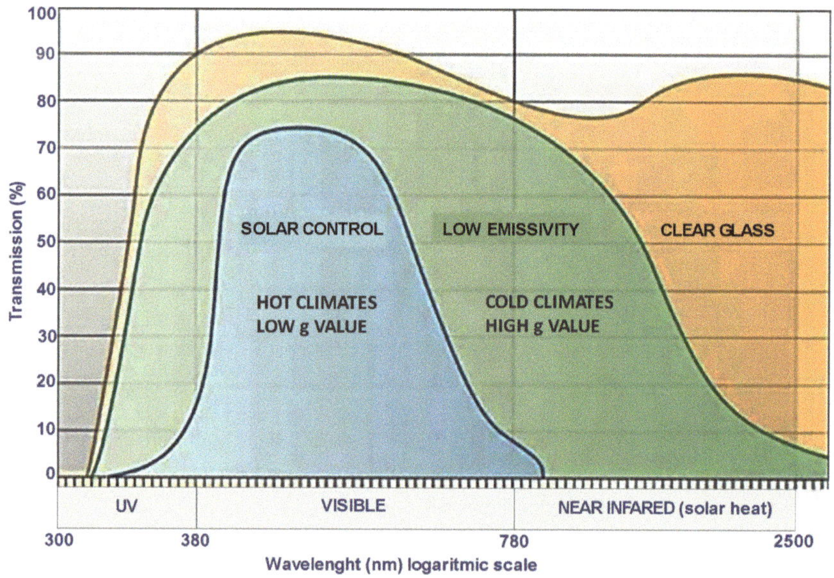

FIG. 1.24 Real coatings transmission areas.

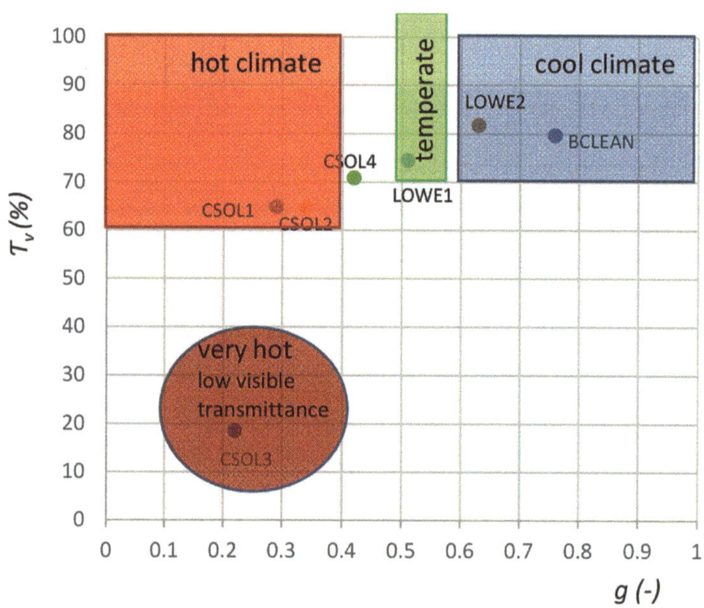

FIG. 1.25 Visible transmittances T_v and solar factors g of seven types of films experimentally compared in the datasheets hereafter.

films actually present in the market and that were experimented by our research group:

LOWEs (1 and 2) are two IGUs with low emissivity films; CSOLs (from 1 to 4) are four solar control films, including one with tinted glass (CSOL3); BCLEAN is one self-cleaning double glass unit. A summary of the results obtained on such coatings is reported in Data Sheets 1.D−1.I.

The graph also reports a hatched area (rectangles/circle) representing ideal couples of T_v and g for different climates.

Deposition methods

Coatings can be divided into two main categories, hard and soft coatings (Jelle, Kalnæs, & Gao, 2015).

Hard coatings are applied during the float line production (online coatings) and are more durable. These coatings are often referred to as pyrolytic coatings. According to the deposition method, doped metal oxide are applied on the glass surface when it is still heated. A widely adopted method is chemical vapor deposition and a largely used type of film is fluoride-doped tin oxide.

As the hard coat becomes as a part of the glass, it exhibits higher durability and can be applied on weather-exposed sides.

Soft coatings are applied to individual glass panes after manufacturing (offline coatings) and require extra protective layers. Therefore, they are also referred to as metal-based multilayer coatings. They are more transparent and have higher infrared reflection, so provide lower emissivity values than hard coatings. The process usually involves high-rate magnetron sputtering with at least one conductive metal layer, such as silver or copper. The most common type is an ultra-thin silver film sandwiched between two dielectric or metal oxide layers (i.e., zinc oxide ZnO, titanium oxide TiO_2, and aluminum-doped zinc oxide AZO).

The metallic layer reflects the thermal and solar radiations. A thin Ag film is the most commonly adopted because of its low value of short-wavelength absorption thus exhibiting higher transmittance in the visible spectrum (Granqvist, 2007). Dielectric or metal oxide layers provide antireflective functions in the visible range and protect the submetallic layer from damage or weathering deterioration. A proper design of layers thickness and the multiplication of layers number determines plural optical interferences, enhancing the visible transmittance while maintaining reflectance to longwave due to the metallic layer. The difference between the refractive index of silver and dielectric layers generates an interference of visible light reducing the visible reflectance and (inversely) increasing the transmittance (Cannavale et al., 2010).

Coatings position

In a standard IGU, coatings can be applied to four potential surfaces. The first (#1) surface faces outdoors, the second (#2) and third (#3) surfaces face each other inside the insulating glass, while the fourth (#4) surface faces directly indoors.

Each coating occupies a specific surface based on its main objective, either to block or to enhance the radiation transmission. Solar control coatings, aimed at blocking the incoming sun radiation, are conveniently positioned in the second surface, while low emissivity films being conceived as outgoing longwave reflecting system, are better to be positioned adjacent to the internal side, namely on surface 3. Some application suggests positioning this latter type also on indoor surface 4 but in this case there would be an increased risk of a loss of condensation resistance.

Low-e coatings

Low-e glazings were introduced in the market in 1979 and have already been adopted as a current standard practice for most European countries for new windows.

The low-e coatings are designed for cold climates to have high values of both solar energy transmittance g and visible transmittance T_v to allow the maximum amount of sunlight into the room.

All materials, including windows, radiate heat in the form of longwave, infrared energy depending on the emissivity and temperature of their surfaces.

Emissivity of the surface of a material is its effectiveness in emitting (i.e., radiating) thermal radiation in the form of longwave, far-infrared energy depending on its temperature. This emission of radiant heat is one of the most important components of heat transfer for a window. Thus reducing the window's emissivity can greatly improve its insulating properties.

The lower the emissivity, the more efficiently the object will reflect infrared radiation (with a mirror-like ability) to keep heat inside the building on cold days. Reducing a material emittance means lowering the radiative heat loss, thus achieving a lower U value and greatly improving the insulating properties (Fig. 1.26).

Standard clear glass has an emittance of 0.84 (Karlsson & Roos, 2001). It means that it absorbs and emits 84% of the radiant thermal energy possible for an object at its temperature and reflects the remaining 16%. Low-e coatings can reach emissivity values even lower than 0.1, which means that 90% of the heat from indoor is reflected back into the room. Very advanced ones could reach emittance values as low as 0.013 (Grynning et al., 2016; Jelle et al., 2015). Such coatings provide very low Ug values, that go down to 1.0 W/m^2 K for a double-glazed unit and

0.5 W/m^{2}K in triple IGUs, thereby contributing significantly to energy saving.

An example of low-e coating applied on surface #3 of an IGU is reported in Fig. 1.27. The values of g and T_V hypothesized in the figure are typical of commercially available products.

Solar control coatings

Solar control glazing are designed for hot climate applications, to reflect NIR radiation (wavelength 0.78–2.5 μm) while allowing the transmittance of visible radiation. They strongly reduce heat gain in summer by blocking a large fraction of the solar infrared energy, thus reducing the cooling loads.

Such type of coatings generally consist of metal films sandwiched between dielectric layers as for low-e coatings. As low-e coatings, solar control ones have low emissivity values to reflect back the longwave radiation and still maintain a low U-factor, but the main difference is in the g value, very lower in the case of solar control films.

Solar control coatings function best when positioned on the glass pane surface closest to the sun, typically the second surface (Fig. 1.28).

In commercial buildings, it is common to apply low-e coatings to both tinted and clear glass. Although the tint lowers the visible transmittance somewhat, it contributes to solar heat gain reduction and glare control.

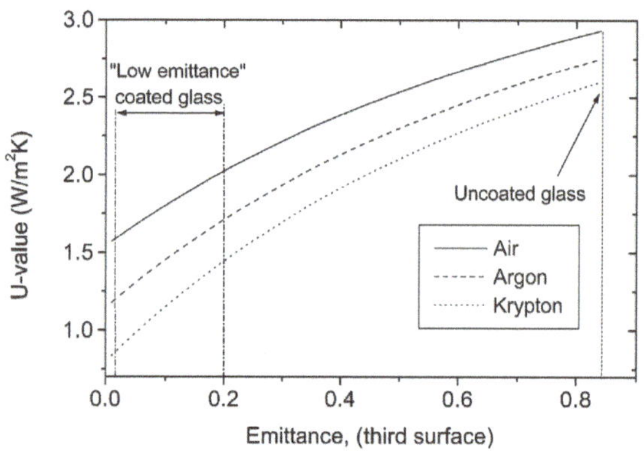

FIG. 1.26 U-value versus emittance for the third surface counted from the outside and for different gas fills. (From Karlsson and Roos (2001).)

Reflective and antireflective coatings

Reflective coatings reduce the solar radiation transmission by increasing the surface reflectivity of the material and thus blocking incoming heat wave. However, the visible light transmittance is very low. They are adopted for external aesthetic uniformity of facades, generally in very hot climates. The reflective coatings are thin metallic films, positioned on the outer face of the IGU (surface #1), so they are more durable than the other films if exposed to aggressive environments.

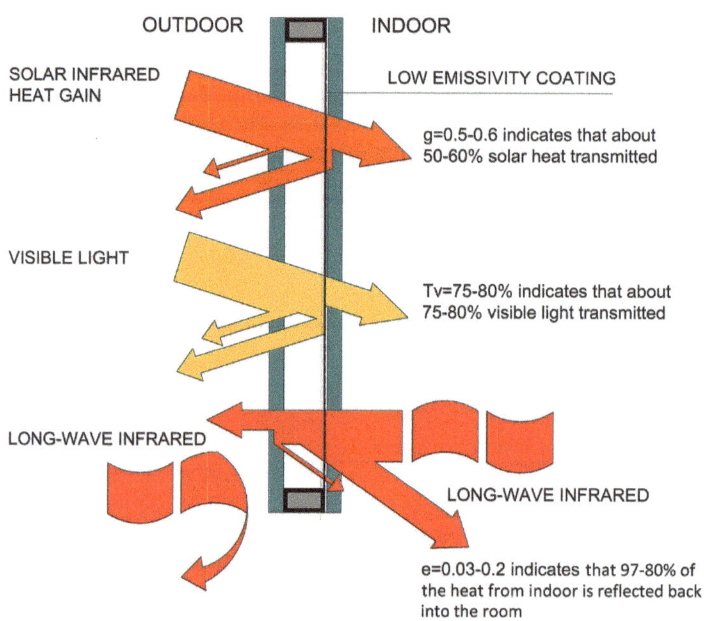

FIG. 1.27 Functioning principle of an IGU with a low-e coating on surface #3.

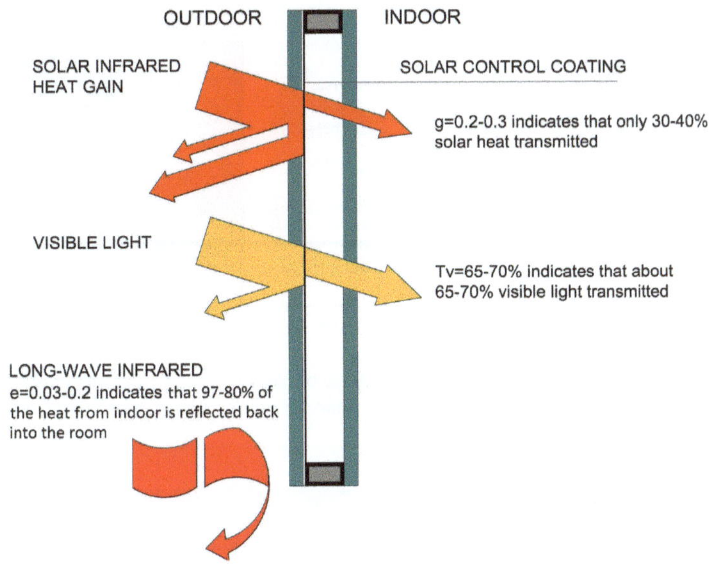

FIG. 1.28 Functioning principle of an IGU with a solar control coating on surface #2.

Adding a low refractive index antireflective (AR) coating on both sides of the low-e coated pane in a double-glazed window makes it possible to achieve high light transmittance while the *U*-value remains unaffected (Rosencrantz et al., 2005). Indeed the antireflective coating aims at increasing the visible transmittance (with enhancements of about 10%) to reduce the electrical lights consumptions rather than at improving thermal performances of the glass. There are two approaches to configure antireflecting coatings: adopting a layer of suited materials with low refractive index (such as porous silica magnesium fluoride or aluminum oxyfluoride), or using a patterned and porous coating (through e.g., nanoimprint lithography).

Such films are broadly used in solar energy applications to enhance the glass transparency.

Tinted glass

In very hot climates, the coatings can be combined with tints.

Using tinted glasses one could find the balance to achieve solar control, help control glare, and contribute to the aesthetics of the building.

Such products are fabricated by adding small metal oxides (to the float or rolled glass composition) that color the glass bronze, green, blue, or gray. Tinted glasses have lower spectral transmittance values for the whole spectrum ranges considered compared to the clear glass of the same thickness. Literature studies have highlighted that blue and green tinted glasses (Fig. 1.29) have the highest visible performance and the lowest thermal transmittance among the existing tinted products in market. They reflect only a small percentage of light so they do not have the mirror-like appearance of glass-reflective coating.

1.D DATA SHEET: LOW E_ HIGH OR LOW SILVER PERCENTAGE?

Several laboratory tests allowed comparing the visual and thermal spectrum of two low-e coatings with different silver percentages. Extended results can be found in (Stazi, Giampaoli, Tittarelli et al., 2016).

Each coating was studied either on a single clear glass (thickness 6 mm) or in surface #3 of a double-glazed unit filled with air (Table 1.3). LOWE1 had a thinner glass on the inner side. Table 1.4 reports the film and glass composition as obtained by X-ray analysis (EDXA) using a high-resolution scanning electron microscope ZEISS SUPRA 40. LOWE1 coating contains a higher share of silver and larger percentages of zinc and titanium with respect to the coating LOWE2.

Composition and Optical Properties

Spectrophotometric tests were done on single glasses with the two coating types to identify the optical performance in visible and solar range. As experimental data revealed, in the initial unaged conditions (Table 1.5 and Fig. 1.30), LOWE1 has the lowest transparency to visible light (τv (%) for single glass, Tv (%) for IGU) and the lowest transmittance even in the

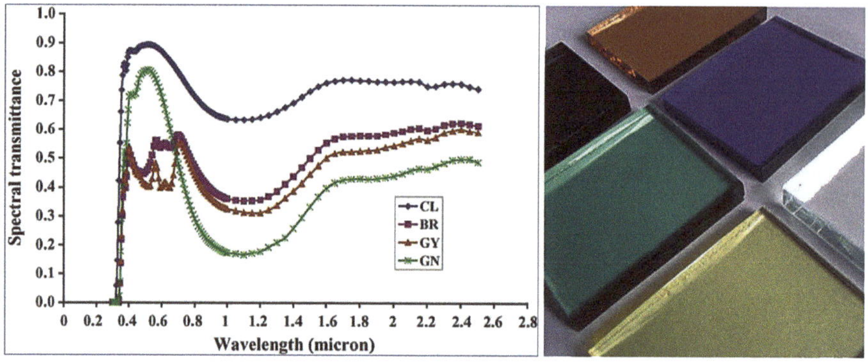

FIG. 1.29 The comparison of the spectral transmittance of clear glass and different tinted glasses of 6 mm thickness: *CL*, clear glass; *BR*, bronze; *GY*, gray; *GN*, green. Experimentation from (Chaiyapinunt et al., 2005).

TABLE 1.3
IGUs Selected for Tests on Low Emissivity Coatings.

Double Glass Samples	Composition (mm)	Gas	Film Position
LOWE1	6/12/4	Air	3
LOWE2	6/12/6	Air	3

TABLE 1.4
Composition of Films and Glass of the Tested Samples With Low-e Coatings.

	FILM COMPOSITION (%)			GLASS COMPOSITION (%)					
Samples	Ag	Zn	Ti	Si	Na	Ca	Al	Mg	K
LOWE1	33.6	14.0	4.2	34.2	6.7	4.1	1.5	1.1	0.6
LOWE2	25.1	7.6	1.8	30.0	13.5	15.0	1.9	3.3	1.8

TABLE 1.5
Optical Properties of Low-e coatings Tested. The T_v, T_e, g and U_g Values Are Obtained for a Double Glass Configuration, Combining the Coated Glass With a Clear One.

	τ_v (%)	τ_e (%)	T_v (%)	T_e (%)	ε (−)	g (−)	Ug (W/(m^{2}K))
LOWE1	80.7	50.4	74.5	46.5	0.033	0.51	1.33
LOWE2	89.3	59.8	81.8	54.4	0.206	0.63	1.71

infrared range (τe (%) for single glass, Te (%) for IGU), indicating a major attitude toward the reduction of the solar gains. Regarding longwave radiation, it shows a very lower ***ε*** value than LOWE2, thus more effectively blocking the outgoing heat flux.

This had a good impact on its U_g value that resulted the lowest one among those achieved in IGUs with the other coatings (see Data Sheet 1.G and 1.L).

Therefore, the highest Ag percentage of LOWE1 determined its better insulation properties.

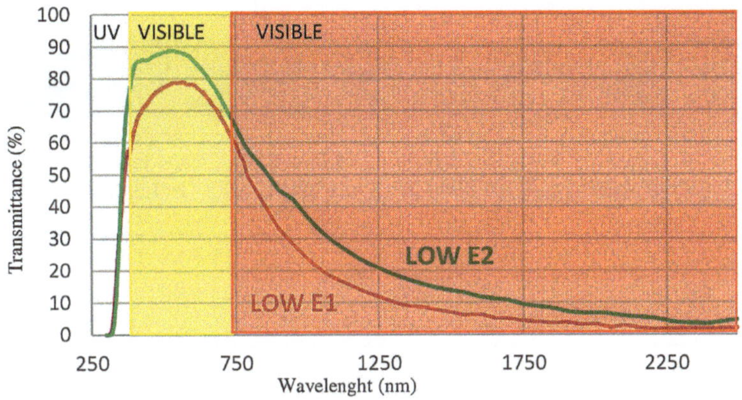

FIG. 1.30 Spectral transmittance of the two low-e coatings tested. (From Stazi, Giampaoli, Tittarelli et al., (2016).)

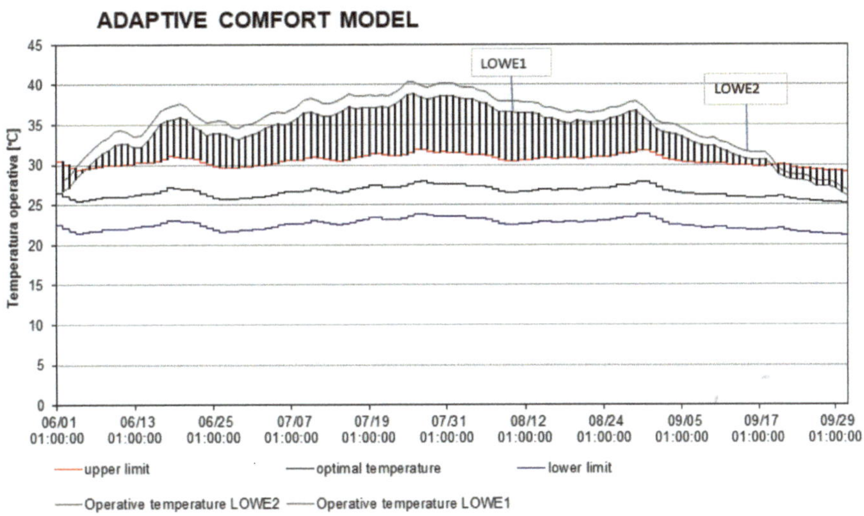

FIG. 1.31 Operative temperatures of two rooms, one characterized by a window with LOWE1 and one with LOWE2.

Summer and Winter Behavior

Experimental measures and numerical simulations in dynamic regime with Energy Plus software on a case study building with low-e coatings (see Section 4.3.1) made it possible to investigate the energy consumptions, comfort levels, solar gains, and daylighting parameters of the two coating types. The comfort level was determined for the summer period with the adaptive model (UNI EN ISO 15251:2008).

As mentioned earlier, low-e coatings are not recommended in hot climates for the recurring overheating phenomena. Having to choose one low-e film in such climates, during the summer the best solutions is that with the highest Ag percentage (33%), namely LOWE1, as it shows lower operative temperatures (Fig. 1.31), consequently higher thermal comfort and lower energy cooling consumptions (Table 1.6).

Regarding the winter consumptions, both films exhibited the lowest energy demand for heating with respect to other films (compare for example the values obtained for the same case study with solar control films in Data Sheet 1.G). The highest Ag percentage of LOWE1 determined its slightly higher performance with respect to the other low-e coating.

Instead, the visual performance was slightly compromised by this higher silver content. Indeed, LOWE1

TABLE 1.6
Energy Saving and Comfort in an Office Room With LOWE1 and LOWE2.

	Solar Gains (kWh/ m^2year)	Discomfort Overheat (hours/season)	Cooling Cons. (kWh/ m^2year)	Heating Cons. (kWh/ m^2year)	Artificial Light Cons. (kWh/ m^2year)	Daylight. Factor (%)	LCA (kg CO_2eq)
LOWE 1	59.5	2530	49.1	32.4	2.7	3.8	4340
LOWE 2	69.1	2599	58.0	32.7	2.6	4.2	4944

shows lower incoming visible light and consequently higher electric energy consumptions to reach proper illumination levels, adopting a minimum value of 400 lux in the working environment.

General Balance

In general, the two coatings showed high g values, around 0.5–0.6, indicating a good transmission of incident solar heat (50%–60%). The visible transmittance was around 75%–80% demonstrating a good natural illumination level. The emissivity is down to 0.033 thus enabling the film to reject indoor the long infrared radiation from internal heat sources.

A general energy balance could be evaluated by considering the global warming potential from LCA study (Table 1.6). It is a relative measure of how much heat a greenhouse gas traps in the atmosphere thus making a global assessment on the environmental impact. The LOWE1 IGU shows the best outcomes.

1.E DATA SHEET: LOW-E IN HOT WET ENVIRONMENT

In low-e films, a thin metal is sandwiched between two dielectric layers to increase the transmittance at visible and near IR wavelengths while decreasing the transmittance at higher wavelengths. Thin Ag films are the most commonly adopted for this core layer.

However, a known problem with silver-based coatings is the formation of white dot defects caused by moisture. Moisture induces silver migration resulting in the peel-off phenomenon of the top oxide layer at the oxide/silver interface. This results in the physical restructuring of the silver and disruption of overlying layers (Ando & Miyazaki, 2001).

Moisture - induced degradation generally occurs when the coated glass substrates are shipped, stored, handled, and cut before final assembly. To address this problem, several approaches are being investigated in the literature: attention during transport and storage, when products are packed with desiccants; perfection of deposition processes; adoption of Ag alloys with Cu, Pd, and Au (Rezaei et al., 2017) for the enhancement of the protective external oxide layers properties; adjunctive encapsulating under and over layers of a particularly chemical resistant material (for example, using a tin silicon oxide or titanium oxide).

Moreover, a characteristic of an Ag-based low-e coating is its instability at high temperatures. In such extreme conditions, the silver could diffuse and agglomerate into island with the consequent decrease of visual transmittance.

In the present section, laboratory tests allowed exploring the durability of Ag-based soft coatings under different aging treatments. The aging under a hot wet environment (HW) provided the insertion of the samples in a thermostatic chamber with 100% RH at 40°C (see Section 4.2.8). Extended results can be found in (Stazi, Giampaoli, Tittarelli, et al., 2016).

Optical and Energy Performance Decay

The graph in Fig. 1.32 and Table 1.7 show that the HW treatment increases the transparency level of LOWE1 shifting upwards its transmittance curve in both visible and solar regions, while decreases the visible and solar transmittance in the case of LOWE2. The causes could be found on the different deterioration modes.

Physical Deterioration

The physical deterioration (Fig. 1.33) was assessed by identifying the percentage of white dots defects after 21 days of exposure. In both coatings the Ag migrated and showed an agglomerating attitude, thus creating islands. LOWE1 had the lowest percentage of white dots, with 18% against 20% of the other coating. The highest silver share of this coating ensured, even after the migration, its highest performance.

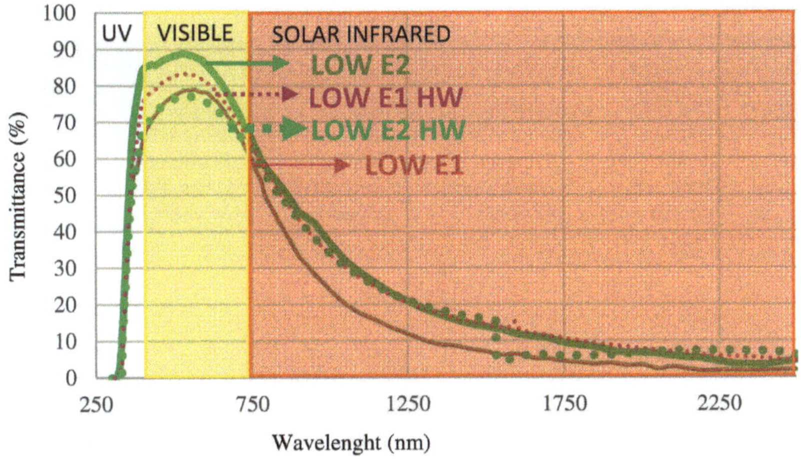

FIG. 1.32 Transmittance of LOWE1 and LOWE2 before and after exposure under hot wet environment. (From Stazi, Giampaoli, Tittarelli, et al., 2016).

TABLE 1.7
Optical Properties of LOWE1 and LOWE2: Parameters Obtained for Unaged (White Lines) and Aged through exposure in Hot Wet Environment (Blue Lines). The T_v, T_e, g and U_g Values Are Obtained for a Double Glass Configuration, Combining the Coated Glass With a Clear one.

	τ_v (%)	τ_e (%)	T_v (%)	T_e (%)	ε (−)	g (−)	Ug (W/(m^{2}K))
LOWE1	80.7	50.4	74.5	46.5	0.033	0.51	1.33
HW exp	82.2	56.4	75.1	51.4	0.18	0.61	1.65
LOWE2	89.3	59.8	81.8	54.4	0.206	0.63	1.71
HW exp	75.9	52.4	69.7	47.8	0.45	0.60	2.18

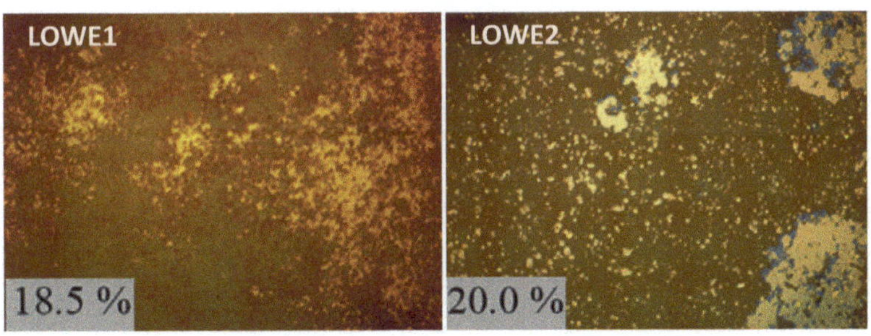

FIG. 1.33 Experimental results on the degradation at day 21.

Final Performance

The degradation determined increased winter heating consumptions for both coatings as a consequence of the increased U_g value (Table 1.8). The increase of g value occurred for LOWE1 lowered its ability of blocking heat gains from the sun, increasing its cooling consumptions or (for a passively cooled room) the discomfort hours for overheating. Instead its increase of visible transmittance has enhanced the daylight illuminance with negligible artificial lighting consumptions decrease.

TABLE 1.8
Energy Saving and Comfort in an Office Room With LOWE1 and LOWE2 Before (White Lines) and After Aging through exposure in Hot Wet Environment (Blue Lines).

	Solar gains (kWh/m^2year)	Discomfort Overheat (hours)	Cooling Cons. (kWh/m^2year)	Heating Cons. (kWh/m^2year)	Artificial Light Cons. (kWh/m^2year)	Daylight. Factor (%)	LCA (kg CO_2eq)
LOWE 1	59.5	2530	49.1	32.4	2.7	3.8	4340
HW exp	65.3	2593	56.9	32.6	2.7	3.9	4873
LOWE 2	69.1	2599	58.0	32.7	2.6	4.2	4944
HW exp	60.6	2481	53.9	36.2	2.8	3.5	4766

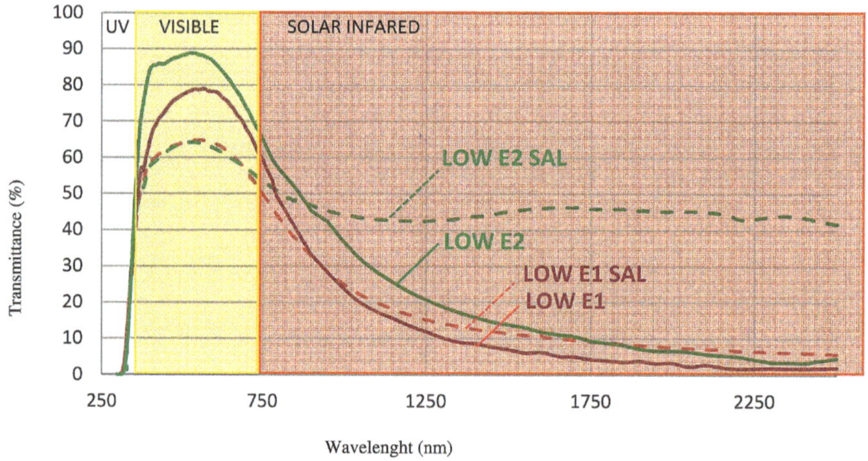

FIG. 1.34 Transmittance of LOWE1 and LOWE2 before and after exposure under saline environment.

General Balance

In general, the two coatings had different behaviors: one increased the entering visible and infrared radiation while the other decreased it.

LOWE1, with the greatest share of silver, had less degraded area, and even after aging conserved the highest silver percentages thus performing well.

1.F DATA SHEET: LOW-E IN SALINE ENVIRONMENT

The aging under saline environment (SAL) consisted in the exposition in a salt-spray machine at 40°C (see Section 4.2.9). Extended results can be found in (Stazi, Giampaoli, Tittarelli, et al., 2016).

Optical and Energy Performance Decay

The graph in Fig. 1.34 shows that the aging treatment has great effect especially on LOWE2 that strongly increased its transmittance for wavelengths outside the visible range. However, the τ_e values are misleadingly reduced (Table 1.9), as this parameter is calculated through a mediated formula between 300 and 2500 nm, also counting low values falling in the visible wavelengths. The great increase of emissivity values ε, also indicates a reduced ability to reflect back the heat (longwave radiation) indoor, especially for LOWE2 case.

Thus, LOWE2 greatly increased the U value. Moreover, both coatings underwent a great reduction of visible transmittance levels.

TABLE 1.9
Optical Properties of LOWE1 and LOWE2: Parameters Obtained Before (White Lines) and After Aging (Blue Lines) through exposure in Saline Environment. The T_v, T_e, g and U_g Values Are Obtained for a Double Glass Configuration, Combining the Coated Glass With a Clear one.

	τ_v (%)	τ_e (%)	T_v (%)	T_e (%)	ε (−)	g (−)	Ug (W/(m^{2}K))
LOWE1	80.7	50.4	74.5	46.5	0.033	0.51	1.33
SAL exp	64.0	44.2	59.3	40.8	0.094	0.49	1.47
LOWE2	89.3	59.8	81.8	54.4	0.206	0.63	1.71
SAL exp	63.3	53.1	58.8	48.3	0.753	0.62	2.7

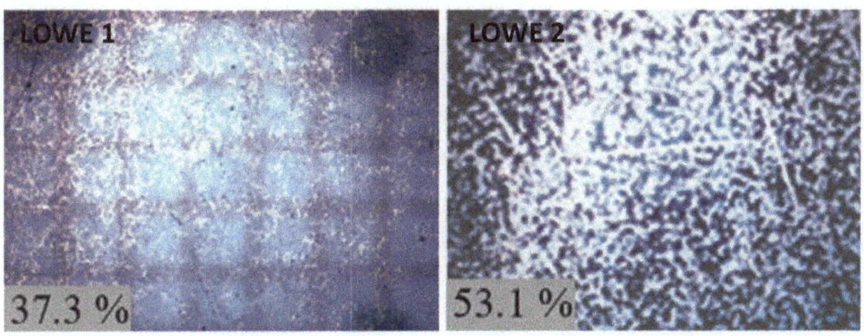

FIG. 1.35 Experimental results on the degradation at day 21.

TABLE 1.10
Composition of Films and Glass of the Tested Samples, Before (White Lines) and After Aging through exposure in Saline Environment (Blue Lines).

	FILM COMPOSITION (%)			GLASS COMPOSITION (%)					
Samples	**Ag**	**Zn**	**Ti**	**Si**	**Na**	**Ca**	**Al**	**Mg**	**K**
LOWE1	33.6	14.0	4.2	34.2	6.7	4.1	1.5	1.1	0.6
SAL exp	24.7	10.1	3.1	44.9	6.6	5.7	2.5	1.4	1.0
LOWE2	25.1	7.6	1.8	30.0	13.5	15.0	1.9	3.3	1.8
SAL exp	12.0	2.6	1.3	37.1	16.9	12.5	5.7	6.5	5.4

Physical/Chemical Deterioration

The study of degradation paths (Fig. 1.35) revealed the lowest percentage of white dots for LOWE1, with 37% against 53% of LOWE2.

The study of the variation on chemical compositions (respect to the unaged coating) using X-ray analysis showed that in both cases the Ag, Zn, and Ti percentages detected are reduced with the simultaneous increase of silica (Si) and sodium (Na) percentages (Table 1.10). This is because where the coating has degradation stains, the silver undergoes a sudden decrease (as migrated away from such zones) and more percentage of the underlying glass substrate components are detected from EDXA.

Final Performance

The degradation determined for both coatings a decrease of visible lighting levels thus determining higher artificial lighting consumptions and decreased visual comfort for minor daylighting factors Table 1.11).

Moreover, the increased thermal transmittance caused an increase of the heating consumptions in particular for LOWE2.

TABLE 1.11
Energy Saving and Comfort in an Office Room With LOWE1 and LOWE2 Before (White Lines) and After Aging through exposure in Saline Environment (Blue Lines).

	Solar Gains (kWh/ m^2year)	Discomfort Overheat (hours)	Cooling Cons. (kWh/ m^2year)	Heating Cons. (kWh/ m^2year)	Artificial Light Cons. (kWh/ m^2year)	Daylight. Factor (%)	LCA (kg CO_2eq)
LOWE 1	59.5	2530	49.1	32.4	2.7	3.8	4340
SAL exp	52.1	2462	48.4	33.4	3.0	2.9	4338
LOWE 2	69.1	2599	58.0	32.7	2.6	4.2	4944
SAL exp	61.2	2434	53.6	38.1	3.0	2.9	4806

TABLE 1.12
Composition of the Tested IGU Samples With Solar Control Coatings.

Double Glass Samples	Composition (mm)	Gas	Film Position
CSOL1	6/12/6	Air	2
CSOL2	6/12/6	Air	2
CSOL3	6/12/6	Air	2
CSOL4	6/12/6	Air	2

TABLE 1.13
Composition of Films and Glass of the Tested Samples With CSOL Coatings.

	FILM COMPOSITION (%)				GLASS COMPOSITION (%)				
Samples	**Ag**	**Zn**	**Nb**	**Sn**	**Si**	**Na**	**Ca**	**Al**	**Mg**
CSOL1	35.2	8.2	–	–	37.7	8.6	6.2	2.2	1.9
CSOL2	24.4	15.1	–	–	39.1	7.6	6.2	4.1	3.4
CSOL3	–	–	33.4	–	52.6	2.4	8.2	2.6	0.8
CSOL4	12.2	10.7	–	29.2	24.5	8.1	7.1	3.6	4.6

General Balance

In general, the two coatings had a similar behavior, with reduced visible light entrance and (especially for LOWE2) a great increase of thermal transmittance. Even under saline environments, LOWE2 performed in a worst way.

1.G DATA SHEET: SOLAR CONTROL—SILVER, NIOBIUM, OR TIN?

Several laboratory tests allowed comparing the visual and thermal spectra of four low-e solar control coatings with different silver percentages and in one case the use of a tinted niobium film.

Each coating was studied either on a single clear glass (thickness 6 mm) or in surface #2 of a double-glazed unit filled with air (Table 1.12). Table 1.13 reports the film and glass composition as obtained by X-ray analysis. CSOL1 contains a higher share of silver than CSOL2; CSOL3 is a niobium-based film; CSOL4 has in addition high Sn percentages.

Other results can be found in (Stazi, Giampaoli, Tittarelli, et al., 2016).

TABLE 1.14
Optical Properties of CSOL Samples. The T_v, T_e, g and U_g Values Are Obtained for a Double Glass Configuration, Combining the Coated Glass With an Uncoated One.

	τ_v (%)	τ_e (%)	T_v (%)	T_e (%)	ε (−)	g (−)	U_g (W/(m²K))
CSOL1	70.3	28.6	64.9	26.7	0.013	0.29	1.58
CSOL2	70.1	33.7	64.8	31.2	0.037	0.34	1.64
CSOL3	20.0	16.4	18.6	15.0	0.647	0.22	2.65
CSOL4	77.2	41.9	70.9	38.4	0.08	0.42	1.74

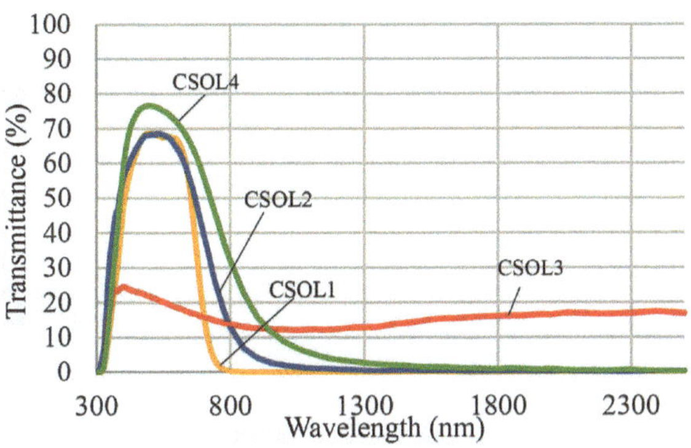

FIG. 1.36 Spectral transmittance of the tested solar control coatings.

Composition and Optical Properties

In the initial unaged conditions (Table 1.14 and Fig. 1.36), the highest Ag within CSOL1 respect to CSOL2 and CSOL4, determined its better insulation properties. Indeed, in the energy range, the elaboration of ε value shows that it has the lowest value of 0.013. Very low g and U values were obtained for this coating in the double glass configuration. CSOL3 has the lowest transparency to visible light for its brown-grey coloring ($\tau v = 20\%$ for single glass, $Tv = 18.6\%$ for IGU). Regarding the longwave radiation, it shows the highest transmittance (highest emissivity value) in far infrared region, thus not effectively blocking the incoming heat flux. However, its τ_e parameter is unexpectedly low as it is calculated through a mediated formula between 300 and 2500 nm, thus including both the visible and the infrared ranges (find the relative formula in Section 4.6).

Summer and Winter Behavior

Numerical simulations in dynamic regime with Energy Plus software on a case study building (see Section 4.3.1) made it possible to investigate the energy consumptions and comfort levels.

As regards the summer behavior, the solar control coatings are recommended in hot climates to avoid overheating. The tinted solution CSOL3 showed that the lowest operative temperatures (Fig. 1.37) but even CSOL1, had a good performance.

Regarding the energy consumptions (Table 1.15), all the films exhibited low energy demand for summer cooling with respect to other film types (compare the values reported for low-e films in Data Sheet 1.D). The highest Ag percentage of CSOL1 determined its highest performance with respect to the other solar control coatings. Instead, the visual performance was greatly compromised in niobium-based film, determining very high electric energy consumptions to reach proper illumination levels, adopting a minimum value of 400 lux in the working environment. Moreover, its heating consumptions were very high.

General Balance

In unaged conditions CSOL1 with the highest silver percentage reveled the best performance, as it showed rather high transmittance in the visible region (300 ÷ 800 nm), and the most convenient curve in the near infrared and in the first part of the medium-infrared range (800 ÷ 2500 nm).

In a general energy balance, considering the global warming potential CSOL1 shows the best outcomes.

1.H DATA SHEET: SOLAR CONTROL IN HOT WET ENVIRONMENT

As low-e coatings, the solar control coatings suffer under high temperature and high regimes of environmental humidity. Soft coatings are generally sealed into IGU and repaired from direct contact with the environment. However, degradation could occur when the coated glass substrates are shipped, stored, and handled before the final IGU assemblies. Moreover, onsite retrofit interventions introduce the biggest

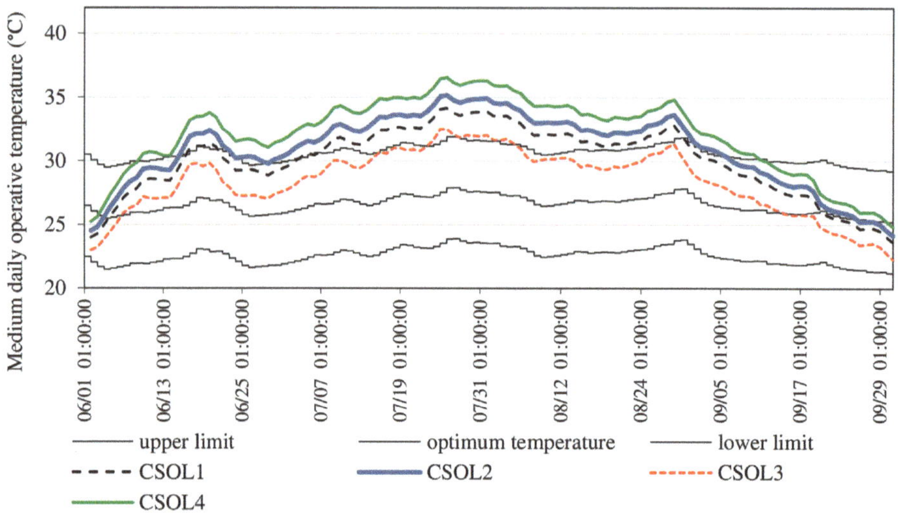

FIG. 1.37 Operative temperatures of office rooms with windows mounting the four solar control coatings under study.

TABLE 1.15
Energy Saving and Comfort in an Office Room With CSOL Coatings.

	Solar Gains (kWh/ m^2year)	Discomfort Overheat (hours/season)	Cooling Cons. (kWh/ m^2year)	Heating Cons. (kWh/ m^2year)	Artificial Light Cons. (kWh/ m^2year)	Daylight. Factor (%)	LCA (kg CO_2eq)
CSOL1	34.2	1429	32.5	37.2	2.8	3.2	3342
CSOL2	39.8	1854	36.3	36.3	2.9	3.2	3583
CSOL3	19.2	400	30.9	44.0	6.5	0.8	3652
CSOL4	48.8	2228	41.7	35.3	2.7	3.5	3911

durability issues. In the present section, laboratory tests allowed to explore the durability of solar control soft coatings under different aging treatments. The aging under a HW provided the insertion of the samples in a thermostatic chamber with 100% RH at 40°C (see Section 4.2.8). Other results can be found in (Stazi, Giampaoli, Tittarelli, et al., 2016).

Optical and Energy Performance Decay

The graph in Fig. 1.38 and Table 1.16 report, respectively, the spectral transmittance curves and the relative parameters for the coated glasses before and after the hot wet treatment. In the figure, only CSOL1 AND CSOL3 curves for unaged conditions are reported (continuous lines) for graph clarity. The unaged curves for the other coatings can be found in the previous section (Data Sheet 1.G).

The results show that the HOT WET treatment decreases the transparency level of CSOL1, CSOL2, and CSOL4 shifting downwards the curves in the visible region. Instead in the solar region the transmittance are slightly increased.

CSOL3, even with negligible modifications, seems to have an inverse behavior with respect to the other coatings.

In general, the HW treatment did not significantly affect the films performances.

Physical Deterioration

The physical deterioration (Fig. 1.39) was assessed by identifying the white dots defects after 21 days of

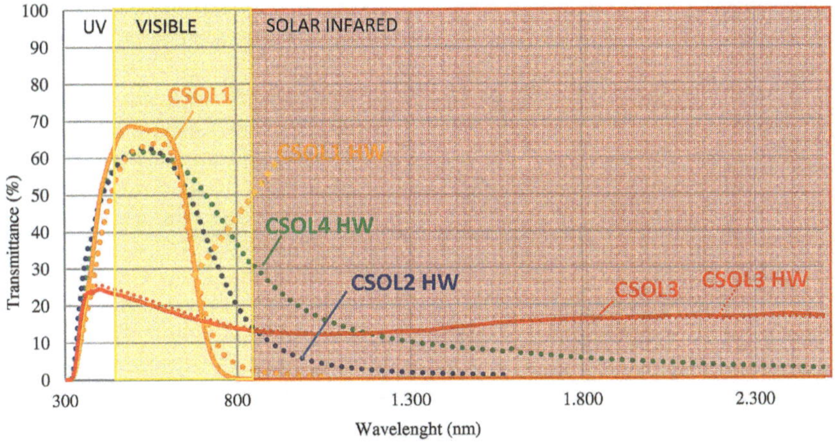

FIG. 1.38 Transmittance of solar control coatings before and after aging in hot wet environment.

TABLE 1.16
Optical Properties of CSOL Samples: Parameters Obtained for Unaged (White Lines) and Aged through exposure in Hot Wet Environment (Blue Lines). The T_v, T_e, g and U_g Are Obtained for a Double Glass Configuration, Combining the Coated Glass With an Uncoated One.

	τ_v (%)	τ_e (%)	T_v (%)	T_e (%)	ε (−)	g (−)	U_g (W/(m^{2}K))
CSOL1	70.3	28.6	64.9	26.7	0.013	0.29	1.58
HW exp	62.2	26.7	57.4	24.8	0.058	0.28	1.69
CSOL2	70.1	33.7	64.8	31.2	0.037	0.34	1.64
HW exp	61.2	32.6	57.2	30.2	0.041	0.33	1.65
CSOL3	20.0	16.4	18.6	15.0	0.647	0.22	2.65
HW exp	21.4	17.9	19.8	16.3	0.601	0.23	2.59
CSOL4	77.2	41.9	70.9	38.4	0.08	0.42	1.74
HW exp	60.9	38.8	55.7	35.5	0.097	0.39	1.78

exposure. CSOL3 had the lowest white dots percentage. Even CSOL2 had a good behavior. CSOL1 with the highest Ag percentage recorded the worst outcome.

Final Performance

The reduction of g value for almost all coatings enhanced their ability of blocking heat gains from the sun, decreasing the cooling consumptions and discomfort for overheating (Table 1.17). The increased U_g value of CSOL1, CSOL2, and CSOL4 led to augmented heating consumptions, while their decreased visible transmittance consequently increased the lighting consumptions.

General Balance

In general, the coatings with and without tinted glass had different behaviors: the former had negligible degradations, while the others experienced reduced visible light and increased infrared radiation transmission. Among all the samples, CSOL1 conserved the best outcome.

1.I DATA SHEET: SOLAR CONTROL IN SALINE ENVIRONMENT

The aging under saline environment (SAL) consisted in the exposition in a salt-spray machine at 40°C

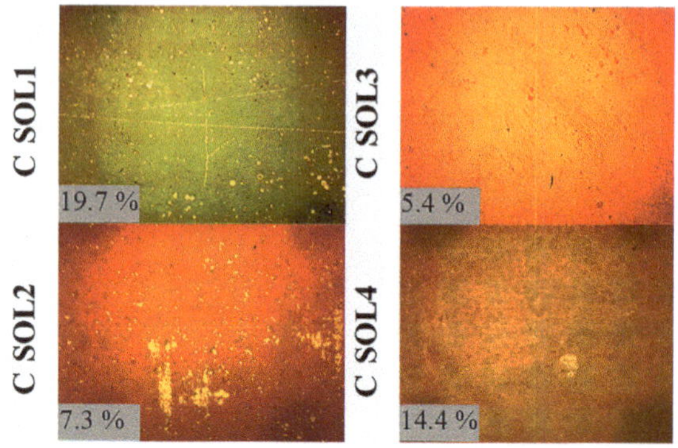

FIG. 1.39 Experimental results on the degradation after 21 days of hot wet exposure.

(see Section 4.2.9). Other results can be found in (Stazi, Giampaoli, Tittarelli, et al., 2016).

Optical and Energy Performance Decay

The graph in Fig. 1.40 and Table 1.18 report, respectively, the spectral transmittance curves and relative optical parameters for the coated single glasses and IGUs after the saline treatment. In the figure, only CSOL1 AND CSOL3 curves for unaged conditions are reported (continuous lines) for graph clarity. The unaged curves for the other coatings can be found above (Data Sheet 1.G).

The results show that the SALINE treatment strongly decreases the transparency level of CSOL1 shifting downwards its curve, while it had lower effects on

TABLE 1.17
Energy Saving and Comfort in an Office Room With LOWE1 and LOWE2 Before (White Lines) and After Hot Wet Aging (Blue Lines).

	Solar Gains (kWh/ m^2year)	Discomfort Overheat (hours/season)	Cooling Cons. (kWh/ m^2year)	Heating cons. (kWh/ m^2year)	Artificial Light Cons. (kWh/ m^2year)	Daylight. Factor (%)	LCA (kg CO_2eq)
CSOL1	34.2	1429	32.5	37.2	2.8	3.2	3342
HW exp	31.9	1123	31.3	38.4	3.1	2.7	3310
CSOL2	39.8	1854	36.3	36.3	2.9	3.2	*3583*
HW exp	38.6	1743	35.4	36.5	3.1	2.7	*3541*
CSOL3	19.2	400	30.9	44.0	6.5	0.8	3652
HW exp	20.7	523	31.6	43.4	6.3	0.8	3671
CSOL4	48.8	2228	41.7	35.3	2.7	3.5	3911
HW exp	45.1	2213	39.8	35.9	3.1	2.7	3824

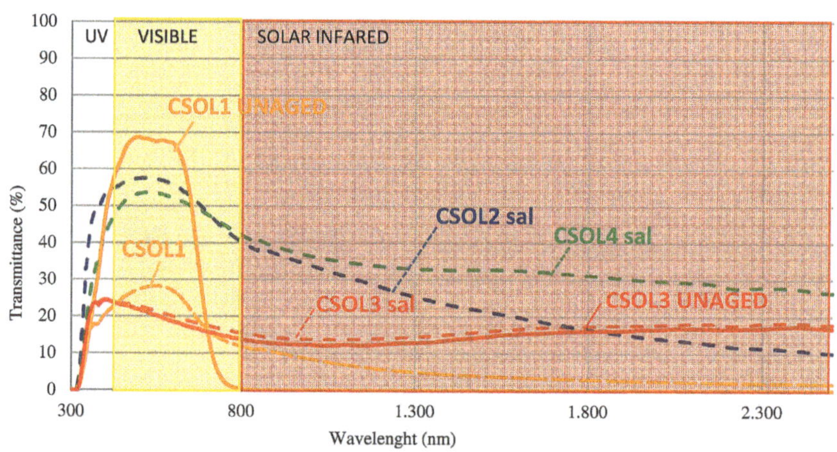

FIG. 1.40 Spectral transmittance of the samples after aging in saline environment.

TABLE 1.18
Optical Properties of CSOL Samples: Parameters Obtained for Unaged (White Lines) and Aged through exposure in Saline Environment (Blue Lines). The T_v, T_e, g and U_g Values Are Obtained for a Double Glass configuration, Combining the Coated Glass With an Uncoated One.

	τ_v (%)	τ_e (%)	T_v (%)	T_e (%)	ε (−)	g (−)	U_g (W/(m²K))
CSOL1	70.3	28.6	64.9	26.7	0.013	0.29	1.58
SAL exp	27.3	16.2	25.5	15.0	0.263	0.20	2.11
CSOL2	70.1	33.7	64.8	31.2	0.037	0.34	1.64
SAL exp	56.6	43.0	52.6	39.4	0.272	0.43	2.13
CSOL3	20.0	16.4	18.6	15.0	0.647	0.22	2.65
SAL exp	22.0	18.7	20.4	17.0	0.622	0.23	2.62
CSOL4	77.2	41.9	70.9	38.4	0.08	0.42	1.74
SAL exp	52.8	42.8	48.7	39.0	0.328	0.43	2.22

visible range on the other films. Instead, CSOL2 and CSOL4 had compromised performance in the solar region, where their transmittances were strongly increased.

In general, the SALINE treatment significantly affected the films performance except for CSOL3 that had negligible degradations.

Physical Deterioration

The physical deterioration (Fig. 1.41) was assessed by identifying the white dots defects. The test consisted in counting the percentage of white spots larger than 0.2 mm. The assessment was done after 21 days of exposure. CSOL4 and CSOL3 had the lowest percentage of white dots, with a value of about 13%–17%. CSOL1 and CSOL2 had the worst behaviors.

Final Performance

The degradation increased winter heating consumptions except for CSOL3 (Table 1.19). This is because of the increase of transmittance levels on regions with high wavelength. Instead the great reduction that occurred on visible region (except for CSOL3) determines strongly lowered visible lighting levels, and a decrease of cooling consumption values or (for a passively cooled room) reduced discomfort hours for overheating.

General Balance

In general, the coatings with and without tinted glass had different behaviors; the former had no degradations, while the others experienced reduced visible light and increased infrared radiation. The global warming potential, obtained from LCA study, revealed that among all the samples, CSOL1 with the highest initial performance conserves the best position.

1.2.3 Functional Design

Multifunctional solutions for glazed surfaces simultaneously optimize several aspects. Among them, the self-cleaning techniques are the most commonly used, for their great potential in durability enhancement and maintenance simplification at low cost.

Self-cleaning coatings

They clean themselves through the action of water and can be divided into two categories, hydrophobic and hydrophilic, according to their different behaviors. The former coatings are usually hazy thus precluding applications on windows, whereas hydrophilic coatings are increasingly used in transparent applications. The self-cleaning behaviour involves a twofold mechanism: many organic molecules as well as microorganisms adsorbed on these surfaces are decomposed under the action of solar irradiation (photo catalysis); when water contacts the surface, it forms sheets and carries away dirt and other contaminants. This latter function imparts to the glass also an antifogging character, as a thin film forms rather than droplets, so the surfaces naturally dry out quickly and prevent fogging up when in contact with condensation.

Moreover, self-cleaning and antireflective properties can be combined to form multifunctional coatings (Ragesh et al., 2014) for photovoltaic applications so as to enhance their efficiency by repelling the dust and dirt molecules and to transmit almost all the light incident on it.

TiO_2 is well suited for application on self-cleaning coatings. It is cheap and readily available, is nontoxic, can be adopted in several morphologies, is easy to deposit, and is chemically inert in the absence of UV light.

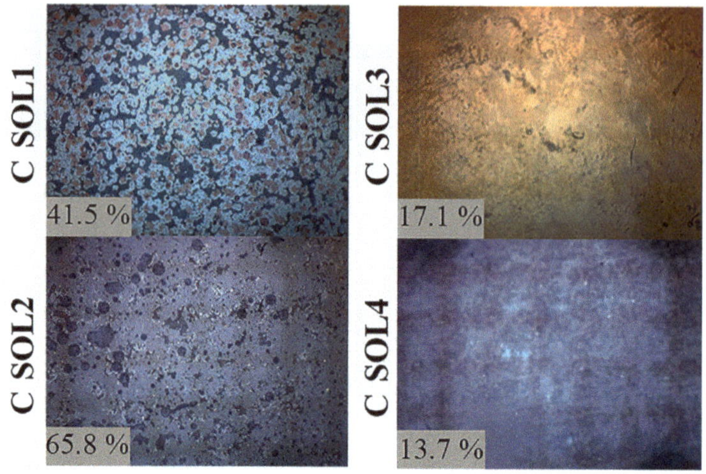

FIG. 1.41 Experimental results on the degradation after 21 days of saline exposure.

TABLE 1.19
Energy Saving and Comfort in an Office Room With CSOL Coatings After Saline Aging.

	Solar Gains (kWh/ m^{2}year)	Discomfort Overheat (hours/season)	Cooling Cons. (kWh/ m^{2}year)	Heating Cons. (kWh/ m^{2}year)	Artificial Light Cons. (kWh/ m^{2}year)	Daylight. Factor (%)	LCA (kg CO_2eq)
CSOL1	34.2	1429	32.5	37.2	2.8	3.2	3342
SAL exp	19.2	221	27.8	42.1	5.4	1.1	3321
CSOL2	39.8	1854	36.3	36.3	2.9	3.2	*3583*
SAL exp	50.3	2098	41.6	37.8	3.2	2.6	*4000*
CSOL3	19.2	400	30.9	44.0	6.5	0.8	3652
SAL exp	21.6	506	31.5	43.6	6.2	0.9	3662
CSOL4	48.8	2228	41.7	35.3	2.7	3.5	3911
SAL exp	49.5	2113	42.5	38.0	3.4	2.3	4072

Nevertheless, it reaches superhydrophilic state upon weak solar irradiation in an ambient atmospheric environment. It has a high refractive index but combined with SiO_2 particles it can also be adopted in antireflective applications (for example, in the surface of solar panels).

As the superhydrophilicity of TiO_2-based films is photoinduced, their performance is gradually lost in the absence of UV light limiting their utility. However, UV-independent superhydrophilicity can be achieved by tailoring TiO_2 morphology. For example, some authors suggest adopting randomly orientated TiO_2 nanofiber mats of various morphologies or nanotube arrays to induce a capillary effect and the consequent superhydrophilicity even in the absence of UV light (Chen et al., 2008; Pan et al., 2007).

1.L DATA SHEET: UNAGED AND AGED SELF-CLEANING COATINGS

Several laboratory tests allowed analyzing the visual and thermal spectrum of a self-cleaning coating. The values were then compared with the low-e and solar control coating discussed earlier.

The coating was studied either on a single clear glass (thickness 4 mm) or in surface #1 of an insulated glazed unit filled with air (Table 1.20). Table 1.21 reports the film and glass composition as obtained by X-ray analysis. Extended results can be found in (Stazi, Giampaoli, Tittarelli, et al., 2016).

TABLE 1.20
IGU Selected for Tests.

Double Glass Samples	Composition (mm)	Gas	Film Position
BCLEAN	4/12/4	Air	1

Composition and Optical Properties Before and After Aging

In the initial unaged conditions (Table 1.22 and Fig. 1.42), the coating led to high transmission values in both visible and solar infrared ranges. The U_g value is the worst if compared with the other experimented coating in previous data sheets (1.D and 1.G) indicating very scarce insulation properties. The ε value is very high if compared with the other films.

The two aging treatments affected the transmittance, with a slight reduction in both visible light and solar range. This could be explained with the photocatalytic effect occurred in the titanium nanoparticles (activated by natural radiation in the aging chambers) that enhanced the film solar reflectance properties.

TABLE 1.21
Composition of Film and Glass of the Tested Samples.

	FILM COMPOSITION (%)	GLASS COMPOSITION (%)					
Samples	**TiO_2**	**Si**	**Na**	**Ca**	**Al**	**Mg**	**Mg**
BCLEAN	1.8	61.7	14.1	15.6	1.0	5.1	0.7

TABLE 1.22
Optical Properties of BCLEAN Samples: Parameters Obtained for Unaged (White Lines) and Aged (Blue Lines) Conditions. The T_v, T_e, g and U_g Values Are Obtained for a Double Glass Configuration, Combining the Coated Glass With an Uncoated One.

	τ_v (%)	τ_e (%)	T_v (%)	T_e (%)	ε (−)	g (−)	U_g (W/(m^{2}K))
BCLEAN	86.9	81.8	**79.7**	**73.5**	0.837	**0.76**	**2.83**
HW exp	77.5	75.6	**71.4**	**68.2**	0.837	**0.71**	**2.83**
SAL exp	80.2	76.0	**74.0**	**68.5**	0.837	**0.71**	**2.83**

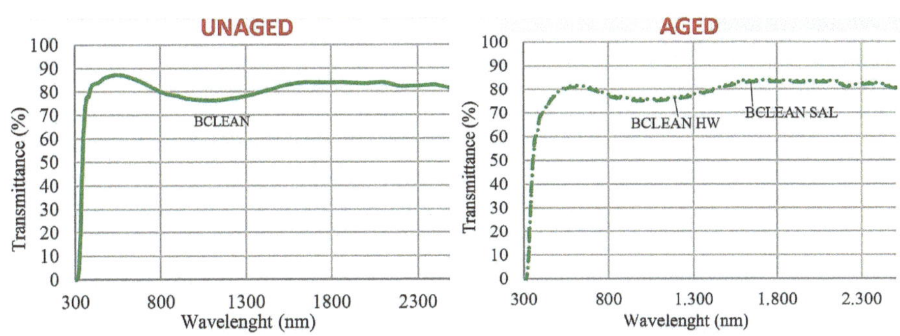

FIG. 1.42 Transmittance of self-cleaning coating before and after the two aging treatments (Stazi, Giampaoli, Tittarelli, et al., 2016).

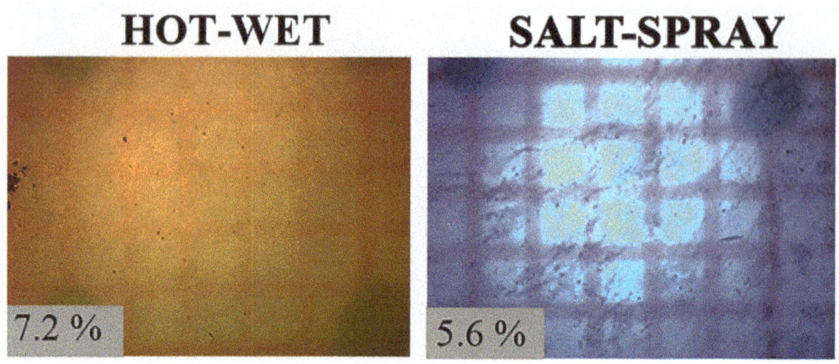

FIG. 1.43 Degradation of the self-cleaning sample after 21 days of saline exposure.

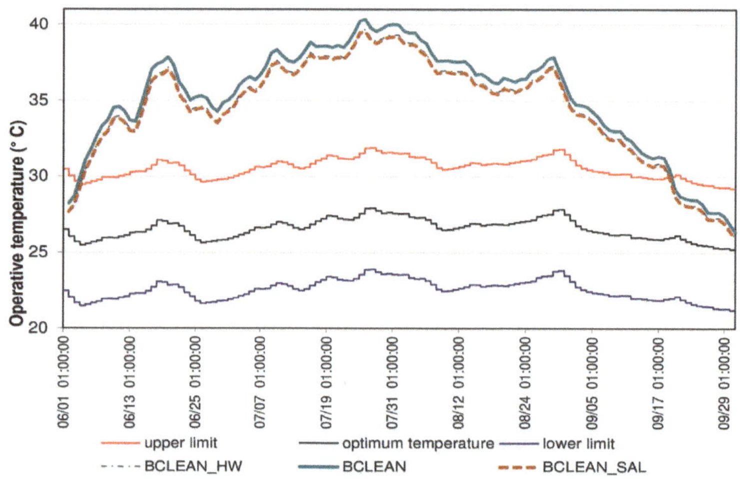

FIG. 1.44 Operative temperatures of office rooms simulated with windows mounting self-cleaning coatings: unaged (BCLEAN) and aged in hot-humid (BCLEAN_HW) or saline (BCLEAN_SAL) environments.

Physical Deterioration

The physical deterioration (Fig. 1.43) was assessed by identifying the percentage of white dots defects after 21 days of exposure. If compared with the other coatings (in previous data sheets), this one shows the highest durability and limited defects after aging.

Summer and Winter Behavior

Numerical simulations in dynamic regime with Energy Plus software on a case study building (see Section 4.3.1) made it possible to investigate the energy consumptions and comfort levels with self-cleaning coatings.

In summer, the room with self-cleaning coating shows overheating (Fig. 1.44). The two aging processes have coincident curves and slightly reduce the temperatures as they activate the photoinduced reflective behavior of the coating.

Regarding the winter consumptions (Table 1.23), the film exhibited rather elevate energy demand for heating with respect to the other film types (solar control and low-e). The visual performance was good, with high daylighting factors.

General Balance

In a general energy balance, considering the global warming potential the environmental impact was high (compared with the other films).

1.2.4 Dynamic Design

The glazing types descripted earlier are all characterized by fixed optical properties, being suitable in either cold or hot climates. However, in most temperate climates, the summertime solar radiation causes unwanted heat loads while winters are cold, so that passive heating by solar radiation is welcome. Research into dynamic glazing systems shows promise of providing variable optical properties, according to the changing of the environmental conditions. The most recent research moves, for example, toward responsive glazing that

TABLE 1.23
Energy Saving and Comfort in an Office Room With Self-cleaning Coatings: Unaged (BCLEAN) and After Hot Wet (HW exp) and Saline (SAL exp) Aging.

	Solar Gains (kWh/ m^2year)	Discomfort Overheat (hours/seas)	Cooling Cons. (kWh/ m^2year)	Heating cons. (kWh/ m^2year)	Artificial Light Cons. (kWh/ m^2year)	Daylight. Factor (%)	LCA (kg CO_2eq)
BCLEAN	96.0	2591	62.9	36.5	2.6	3.9	5369.8
HW exp	90.3	2562	59.8	37.1	2.7	4.1	5181.5
SAL exp	89.5	2550	59.3	37.3	2.7	4.0	5152.6

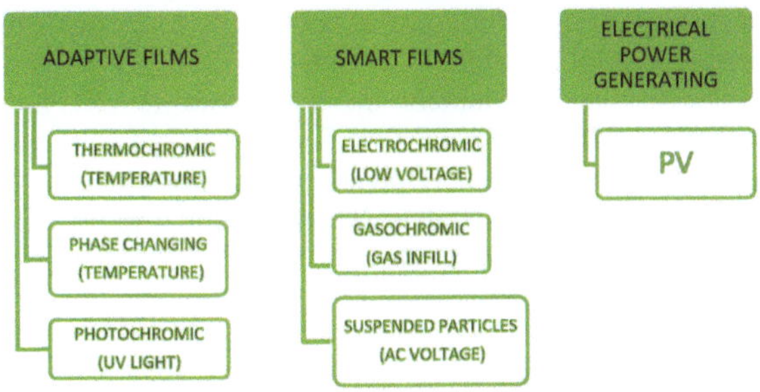

FIG. 1.45 Dynamic coating types.

can change the transmittance depending on available daylight (Rezaei et al., 2017).

Dynamic coatings (Fig. 1.45) could respond either to specific environmental stimuli, in this case usually named "adaptive coatings", or to an external energy source, in this case referred to as "smart coatings." Therefore, only the latter allow a direct control by the users according to their needs. These responsive behaviors are operated either by internal physical property changes (passively) or through external energy exchanges (actively) (Velikov & Thun, 2013).

ADAPTIVE MATERIALS have physical properties and structures that passively adapt to real time environmental changes. So they can be stimulated by a change in temperature, light, humidity, or carbon dioxide. Thermochromic (activated by temperature), photochromic (activated by ultraviolet light), phase-change materials (temperature reactive), and Tio_2-based materials (activated by sunlight) can be classified as adaptive solutions.

SMART MATERIALS differ from adaptive materials, as they require external stimuli based on conventional energy sources, such as voltage, light, or heat (Al-Obaidi et al., 2017). Some examples are electrochromic, gasochromic and suspended particle devices.

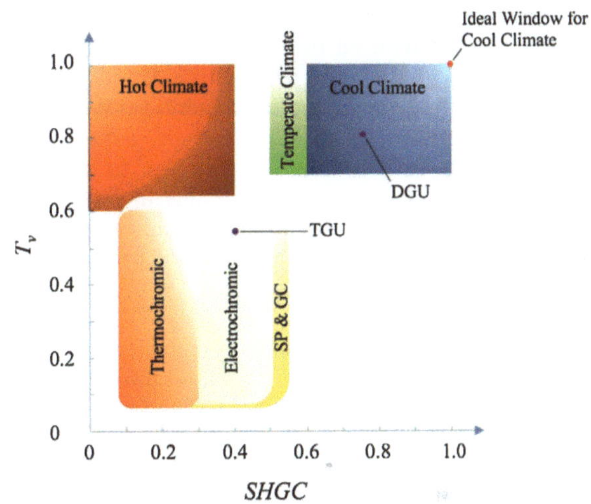

FIG. 1.46 Dynamic coatings and climates: SP = Suspended Particles; GC = Gasochromic; DG = Double Glazed Unit; TGU = Triple Glazed Unit; (Rezaei et al., 2017).

ELECTRIC POWER GENERATING COATINGS (i.e., photovoltaic transparent modules) can be incorporated within these latter energy-needing techniques to make them self-sufficient.

The previously mentioned dynamic coatings are mainly for hot and temperate climates (Fig. 1.46). With visible transmittance enhancements, thermochromic coatings (TC) would fit for hot climates, while electrochromic (EC), suspended particle (SP), and gasochromic (GC) windows would satisfy minimum requirements for hot-temperate climates. Moreover, the increase of SHGC (namely the Solar Heat Gain Coefficient, corresponding to the Total solar energy transmittance, g) and the adoption of these technologies into multipane windows would make it possible to achieve the required U-value for cold climates as well.

The subsequent text reports a general overview on the most common and recent dynamic techniques. Other information could be found in (Casini, 2018).

Adaptive films

Thermochromic coatings. They are temperature reactive. They vary the solar IR radiation transmittance with temperature variation, being opaque at a critical temperature and returning transparent when the temperature drops. The most widely studied thermochromic oxide material is VO_2, which exhibits a structural transition at a "critical" temperature below which thin films are semiconducting and relatively nonabsorbing in the infrared and above which they are metallic and infrared reflecting.

However, for practical uses, the transition temperature should be reduced from the bulk value of 68°C to values close to the room temperature. This drawback could be solved through doping elements (as W and F). Another disadvantage of this type of glass is the nongradual transition from transparency to opacity, thus decreasing the passive solar gains in the winter period. Moreover, the visible light transmittance is low so the combination with antireflective coatings could be useful to increase its value.

Phase changing devices. They are temperature reactive. They can be used directly within the IGU cavity or to fill containers placed between window glass panes. They change phase from solid state to liquid as absorbing energy from the heat source. Therefore, they absorb infrared radiation. When the temperature drops, the liquid state turns into solid state by releasing the previously stored energy. Paraffin is the widely used PCM.

Their disadvantages are the poor visual transmittance, the low thermal conductivity, and the large volume change, aspects that limit their application in buildings (Cuce & Riffat, 2015).

Photochromic coatings. They are light reactive. They allow a reversible change of color upon exposure to ultraviolet light and visible light. So they become optical absorbing in the full wavelength range of solar energy. The material then changes to colorless when removed from the UV light source.

Their disadvantage is the scarce visible light transmission, which turns out in higher heating consumptions especially in cold climates.

Smart films

Electrochromic devices. The external stimulus is the electricity (DC voltage). This technology is based on the tendency of a number of solids, such as the oxides of tungsten, vanadium, molybdenum, or titanium, to perform a reversible color change when subjected to an electric current or field. These materials may be suitable for use in doping sol–gel glass to create a glazing system that can alter its transmissivity in response to an electric signal. However, research into electrochromic windows is still at an early stage.

Gasochromic devices. The external stimulus is the hydrogen gas. In this technology, a tint alteration is achieved by exposing a gas sensible layer to diluted hydrogen (H_2) or oxygen (O_2) pumped in the window cavity. Glass surface #2 is coated with WO_3 and an extremely thin surface layer of catalytic platinum. The coated side of the glass is in contact with a gas that is confined in the IGU. Changing the amount of H_2 in contact with the thin film, makes it possible to incorporate variable amounts of H^+. The catalyst undergoes a chemical reduction and darkens and the transmittance is then modulated. As hydrogen is explosive in air over 4% concentration, for windows operation it is usually diluted in an argon carrier to stabilize the compound and improve the thermal insulation.

Suspended particles devices. The external stimulus is electricity (AC voltage). In such devices, dipolar particles are suspended within a water-like organic liquid between two indium tin oxide (ITO) layers. In the opaque state, the randomly oriented particles absorb light. When a voltage is applied to the SP window, the randomly oriented particles align with the electrical field and the window become transparent. Different transparency levels can be reached by controlling the choice of absorptive particles, the particle density, and the liquid layer thickness. SP devices have response time considerably faster than EC windows but require higher energy consumptions.

Moreover, as they are dark in their off state, they are more suited to vehicles (in that darkened glass is desirable when they are parked) rather than to buildings.

Photovoltaic glazing

Encapsulated glass-to-glass photovoltaic modules with antireflective and photocatalytic coatings are a promising technology, combining renewable power generating and the destruction of air pollutants in the urban environment. However, since windows need transparent state, such systems cannot produce electricity from the visible part of the solar spectrum. To address this challenge new solutions for transparent cells are continuously being developed, including thin film technologies, dye sensitized solar cells, organic molecules carrying the re-emitted infrared light into transparent solar concentrators.

TABLE 1.24
Uw Values for IGUs With Different Characteristics.

Uw (W/m^2K)	Warm Edge Spacer Thermal Break Frame	Aluminum Spacer Nonthermally Broken
Triple pane low-e coated argon-filled $U_{cog} = 0.24$	1.7	2.8
Double pane low e coated air-filled $U_{cog} = 0.29$	1.9	3

1.3 FRAMES

The frames have great importance on the overall performance of the glazed system. Indeed, although window frames represent 20%–30% of the overall window area, their impact on the total window heat transfer may be much larger, especially when combined with very low conductance glazing. Today, very highly insulating frames are being developed in the market. U_f values typically vary between 0.8 W/m^2K and 2.4 W/m^2K but the frames recognized as highly efficient in the market all have U_f values lower than 0.7 W/m^2K (Jelle et al., 2012). Some commercial product reaches U_f value as low as 0.5 W/m^2K (Van Den Bossche, Buffel, & Janssens, 2015).

Limited information on thermal optimization of frames are available in the literature, and it is clear that frame U values are still far from reaching their glazing counterpart so the decrease of U_f value should be an industry priority. The values reported in the literature allow defining the scheme in Table 1.24.

The data show that the frame and edge of glass control the overall U value. Indeed, the most impactful change is the switching from nonthermally broken to the thermally broken frame and the replacement of the highly conductive aluminum spacer within the IGU with a low conductance warm edge spacer. Therefore, this stresses the importance of improving the performance of frame and edge of glass before the maximum enhancement of the glass packages.

1.3.1 Materials and Components

Traditionally, the frames are conceived in wood, thermoplastic (e.g., vinyl), and aluminum or a combination of them. For example, clad wood is wood that has been coated on the exterior with insulation-filled aluminum cladding.

The traditional frames with current design and materials difficultly reach the increasingly strict energy efficient standards. Although new materials for frames have been developed, the cost of retooling has sometimes discouraged their adoption and the manufactures often rely on increased insulations, coated glass, and multiple glazing. Moreover, the position in which the frame is placed with respect to the wall is very important, making possible to reduce the thermal bridge losses by more than 50% (Misiopecki et al., 2018).

Fig. 1.47 reports the generic window frames for the most commonly used profiles in aluminum, wood, and vinyl. Table 1.25 reports the U_{frame} (U_f) values for these typical frames configurations.

Aluminum

Currently, commercial windows mostly employ aluminum because of its relatively low cost, high strength, easy manufacturability, and long service life. Such material has been used for many years for its high strength in windy areas and tall buildings. It is far stronger than vinyl and wood and has very higher elasticity modulus than such other two materials. Its main issue is the excessive transmission of heat (high U values) so it achieves the low energy standard values with greater difficulties than the other materials. Moreover, it has medium-high levels of thermal expansion.

The U_f value of a double glazed window with a typical aluminum frame 70 mm thick is about 2.7 W/m^2K. This value could be halved (apart from multiple glazing options) by subdividing the thermal break in separate cavities and filling the cavities between IGU and frame with insulation material (Van Den Bossche et al., 2015).

The main elements of an aluminum-framed window are visible in Fig. 1.47A, namely the master fixed frame and the upper operable sash.

There are several window weather-strips available on the market with different material, size, thickness, and support configuration. They run continuously between sash and frame elements of a window to repel air infiltration and water penetration.

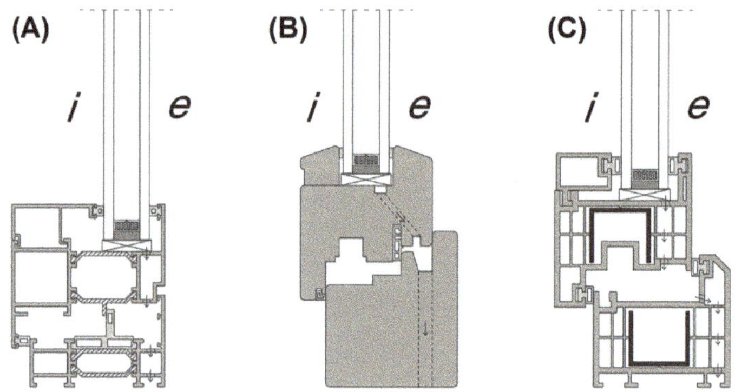

FIG. 1.47 Typical window frames in aluminum **(A)**, wood **(B)**, and vinyl **(C)**. (From Van Den Bossche and Janssens (2016).)

TABLE 1.25
Uf Values of Most Used Framing Systems (Van Den Bossche et al., 2015).

U_f values(W/m^{2}K)	Double Glazing 24 mm	Triple Glazing 42 mm
aluminum	2.773	2.618
wood	1.707	1.640
vinyl	1.503	1.451

The windows commonly utilizes compression weather-stripping (also named gaskets) that, as compressed between sash and frame in the closed position, seal better than sweep-type strips.

With regard to the profiles design, very important has been the transition from "face seal" configuration that featured a closed joint toward the external environment to the "rain-screen" configuration characterized by the adoption of an open joint.

In the former case, the frame is a perfect barrier respect to the external weather, relying on tightly sealed joints using caulking and/or gaskets. The water-shedding surface is coincident with the exterior moisture barrier and air barrier (Fig. 1.48A).

In the latter case, the frame adopts the rain screen principle to prevent the penetration of running or wind driven rainwater. It envisages the presence of a cavity through which water may drain to the exterior and ensures the pressure equalization, so that rainwater will not be drawn indoor (Fig. 1.48B).

Another important step in aluminum frames technological evolution was the adoption of thermal break profiles to avoid the heat dispersion. Technologies commonly used for thermal break can be divided into two categories:

1. Pour and debridge method (Fig. 1.49).

An aluminum profile is designed and extruded with a cavity to limit the bridge of metal. The cavity surface is abraded with a conditioning primer to mechanically lock the polyurethane polymer with the finished surface of the aluminum profile. A two-component polyurethane dispersed as a liquid into the thermal barrier channel solidifies into a very strong low conductance structural polymer. The final step is the removal of the metal bridge from the bottom of the channel. This older method is still in widespread use. Windows incorporating this type of thermal break generally reach R2 thermal performance.

2. Use of separating polyamide strips or composite elements (Fig. 1.50).

The polyamide strips are sometimes called I-bars. The frame is extruded into two dies, and polyamide strips (usually two) are crimped on each side to create single framing cross-section. Even though polyamide has higher conductivity than polyurethane, the strips are thinner and can reach larger widths than pour and debridge systems. Windows incorporating this type of thermal break generally reach R3 thermal performance.

The composite elements are instead realized through polyurethane foam. A thin layer of aluminum cladding covers solid polyurethane elements, minimizing direct connections between inside and outside.

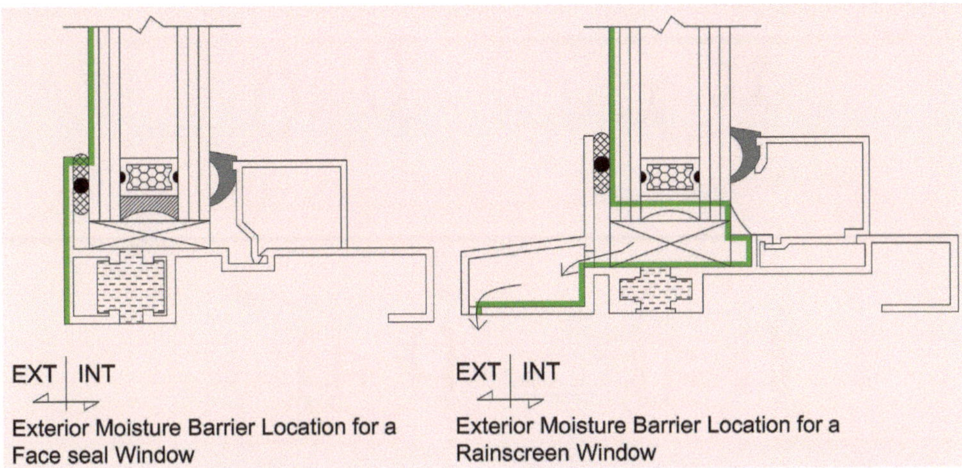

FIG. 1.48 Two different types of water resistance methods, in the evolution of the frame design: **(A)** face seal; **(B)** rain screen. From: Water penetration resistance of windows—study of manufacturing, building design, installation, and maintenance factors. (Source: Canada Mortgage and Housing Corporation (CMHC).)

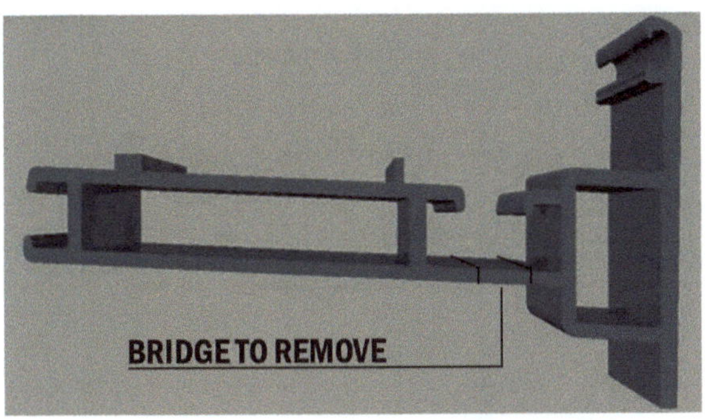

FIG. 1.49 Pour and debridge method.

The historic evolution of the profiles was toward increasingly complex polyamide bars design (Fig. 1.51). Moreover, the sealing strips evolved from simply to complex multicamera elements (Fig. 1.52).

Such complexifications in the historic evolution of the frames determined a strong enhancement of their thermal properties (Fig. 1.53).

Wood

The U_f value of a double-glazed window with a typical wooden frame (hard wood) 70 mm thick is about 1.7 W/m^{2}K. However, it is possible to achieve reduction of 64% by choosing the right combination of optimization techniques (Van Den Bossche et al., 2015). Possible thermal improvements (apart from multiple glazing options) regard the adoption of thicker sections, use of soft wood (but with higher durability problems), or choosing a combination of wood and other materials (such as cork).

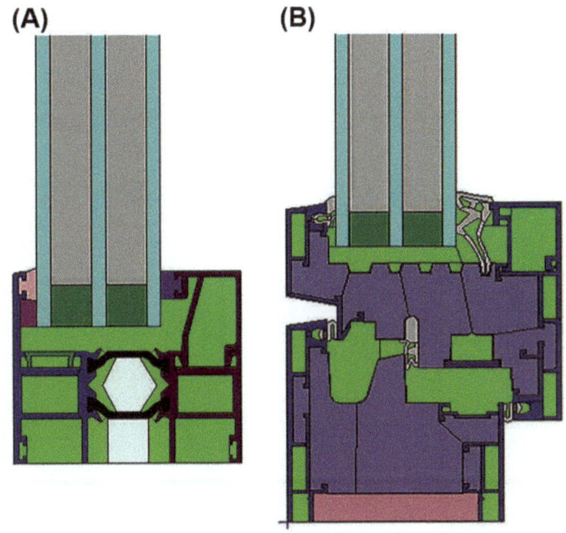

FIG. 1.50 Typical thermally broken aluminum frames, from (Gustavsen et al., 2011): **(A)** with polyamide thermal breaks; **(B)** with polyurethane (PUR) foam insulation elements.

The main issue of wooden frames regard the high maintenance requirements.

Vinyl

The U_f value of a double-glazed window with a typical vinyl frame 70 mm thick is about 1.5 W/m^{2}K. The thermal resistance could be improved (apart from multiple glazing options) up to 50% by deepening

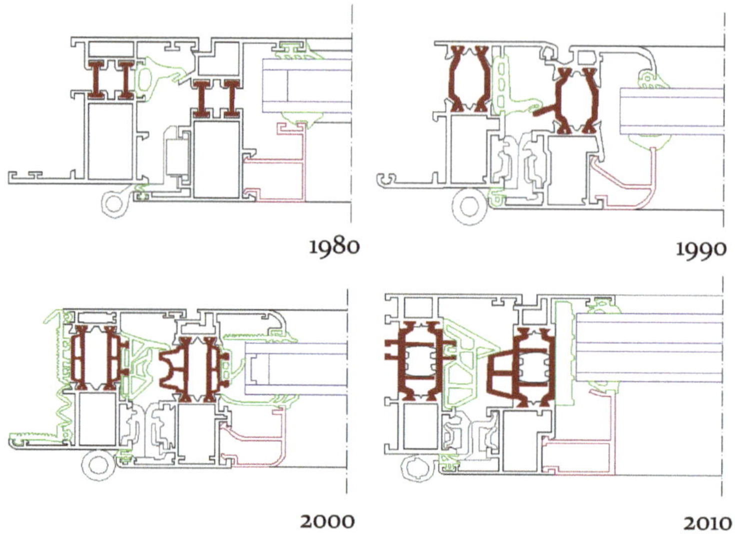

FIG. 1.51 Evolution in the design of polyamide strips.

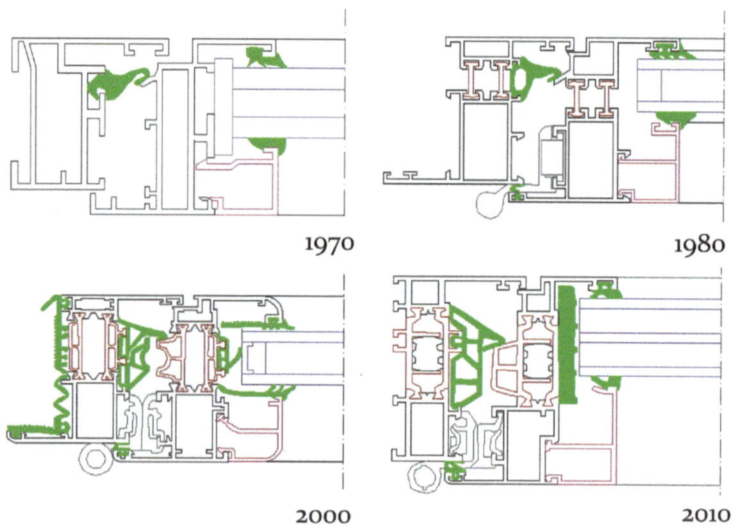

FIG. 1.52 Evolution in the design of sealing strips.

the frame up to 120 mm and inserting insulating material in substitution of the steel reinforcement (Van Den Bossche et al., 2015).

The main problem of such type of frames is the lack of ability to withstand weather extremes. In particular, PVC has a thermal expansion coefficient very higher than aluminum and 14 times higher than glass. So it expands and contracts up to 14 times higher than such material and loses its ability to support heavy loads at the temperature rising. Another weak point of such frame is the necessity to insert a steel reinforcement. Such reinforcement could introduce thermal bridges.

1.3.2 Emerging Technologies

The steel reinforcement needed for vinyl frames is instead unnecessary by adopting a section in glass fiber reinforced polymer (GFRP) that would even more decrease the thermal transmittance, but at higher costs and with other issues regarding manufacturing and durability (Stazi, Giampaoli, Nisi, et al., 2016). The most advanced frames of such materials are able to reach a 0.20 U factor.

Every type of window (single hung, double hung, vertical, horizontal sliders or fixed) requires specific frame-sash details, thus separately designed ad hoc

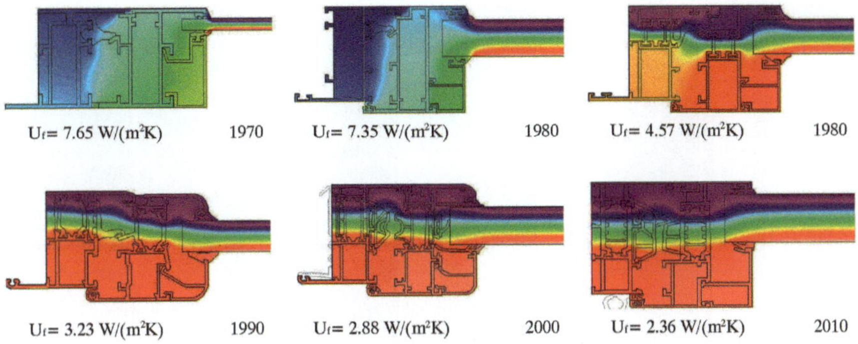

FIG. 1.53 Evolution of the thermal performance of the aluminum frames.

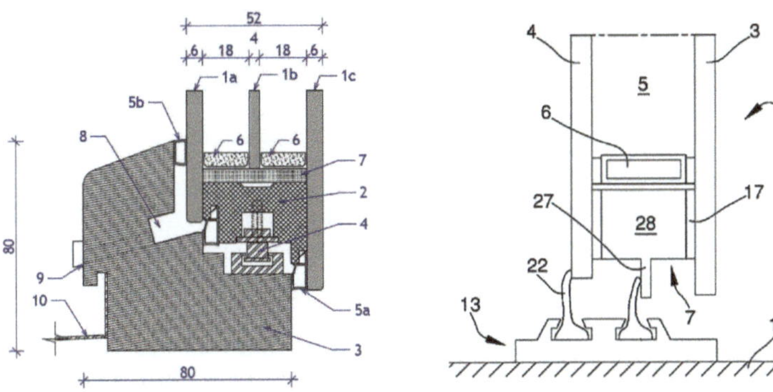

FIG. 1.54 Integrate framing (Oravec, 2015); Invisible window frame (Patent EP 3 071 775 B1).

extrusions and relative tools. A complete window system typically requires 16–17 extrusions. Therefore, it is very difficult to explore new materials and designs for high costs and manufactures reluctance.

Great innovations are recently regarding extrusion techniques (patent US 2018/0038144 A1), integrated framings adopting the most energy efficient materials (Fig. 1.54A), namely wood and composite (Oravec, 2015) or very thin structural framing inserted within the IGU (Fig. 1.54B), that relies on the structural continuity along the entire perimeter of the glass to favor a ring-like behavior (patent EP 3 071 775 B1).

1.4 SHADINGS

The present section is about the shadings technologies. It begins with static solutions, and then addresses dynamic and highly advanced ones.

The performance of the combined shading-glazing system depends on various parameters, namely climate, glazed area (small or large windows), glazing type (U value and total solar transmittance factor), and orientation. Moreover, the final performance strongly depends on shading device (visual and thermal) properties and on its position with respect to the glass. So the perfect solution should be identified according to a case-by-case evaluation (Atzeri, Cappelletti, & Gasparella, 2014).

1.4.1 Static Design

Internal blinds

The greatest part of the existing literature focuses on internal blinds, by comparing the best performance among several configurations, such as curtains, venetian blinds, polyester blinds, PVC blinds, and fabric roller shades.

Even in very cold conditions (Bessoudo et al., 2010), the interior glass surface temperatures can be quite high during sunny days, resulting in thermal discomfort for excessive operative temperature and radiant temperature asymmetry. Therefore, the use of an internal shading even on clear winter days of the coldest climates, can improve the indoor thermal environment by minimizing the amount of direct solar radiation.

In general, the experimental studies in literature (Fig. 1.55) agree on the fact that the most influencing features on energy performance are the combination

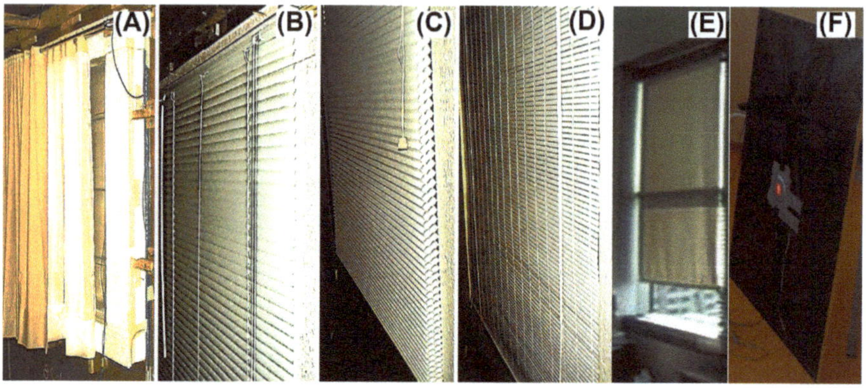

FIG. 1.55 Experimentations on internal blinds: **(A)** double curtain, **(B)** venetian blind, **(C)** polyester blind, **(D)** PVC blind inside the window frame, **(E)** roller shade, and **(F)** low-e shade. Figures **(A–D)** regard experimentations from (Cuevas, Fissore, & Fonseca, 2010).

of shade absorptance and glazing properties, the distance of the shade with respect to the glass and its permeability with respect to the airflow.

When the shading is very close to the window and not permeable, a cavity effect is produced thus determining turbulences inside the confined space and therefore increasing the heat transfer coefficient. Moreover, for slat types, the performance depends on the tilt angle of the slats. Maintaining slat angles at a horizontal position can still cause discomfort for the large amount of the incident solar radiation. A tilt of 45 degrees will decrease the amount of radiation entering in the space.

Shadings within double skin facades

Different types of shadings can be inserted within the cavity of a double skin façade. For example, one interesting experimental study (Serra, Zanghirella, & Perino, 2010) compared aluminum venetian blind and PVC-reflecting rolling screens. In this case, especially for cold climates, the highest performance could be reached by selecting the system with the greatest ability to preheat the ventilation air. Therefore, venetian blinds resulted better than reflecting roller screens.

Inversely in summer, and hot climates, the use of a reflecting roller screen always leads to higher efficiency values, with respect to the venetian blind. In the former case, the air cavity is divided into two different airflow channels. The external one collects most of the radiation absorbed and reflected by the shading device, and the air in the internal one has a more effective insulating and cooling effect. This does not occur with the venetian blind, whose higher air permeability determines a higher mixing of the air in the cavity.

Compared with a traditional insulated glass unit (IGU), both systems were found to be effective regarding energy efficiency and thermal comfort performance issues. Nevertheless, they are still not competitive with opaque facades and need further improvement to avoid overheating in cooling periods and to enhance the ability to behave as a solar collector in the heating periods. Actually, systems based on phase change materials and within multiple glazing double skin facades are under test (Fig. 1.56).

External shadings

It is well established that the external shadings allow reaching higher summer performance with respect to internal solutions (Ylitalo, Ip, & Marshall, 2006). However, comparative experimental studies on the external shading devices through simultaneous measures are limited, being mainly the assessment of one single device or numerical comparisons. External solutions include roller shades, external binds, external panels (also with low-e properties), and external louver devices.

1.M DATA SHEET: UNSHADED VERSUS INTERNAL LOW-E SHADING

An experiment was carried out on a school building by measuring the performance of unshaded glass in comparison with an internally shaded glass. The latter involved the use of an internal low emissivity shade (Fig. 1.55F). Details on the case study are reported in Fig. 1.57 and in Section 4.3.2. In a preliminary experiment, the shade was mounted on one window casing, leaving the other casing unshaded, to compare with the same indoor conditions the surface temperature values. Subsequently, two adjacent classrooms on the Southern side were selected, and the shading system was mounted on the internal window side on only one of them. Extended results can also be found in (Ursini Casalena, 2012).

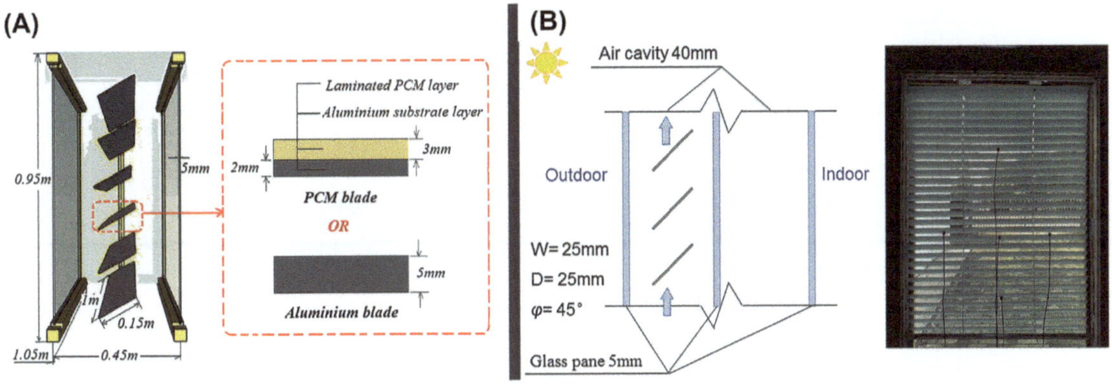

FIG. 1.56 Examples of shading systems within a double skin: **(A)** from (Li, Darkwa, & Kokogiannakis, 2017); **(B)** from (Zhang et al., 2018).

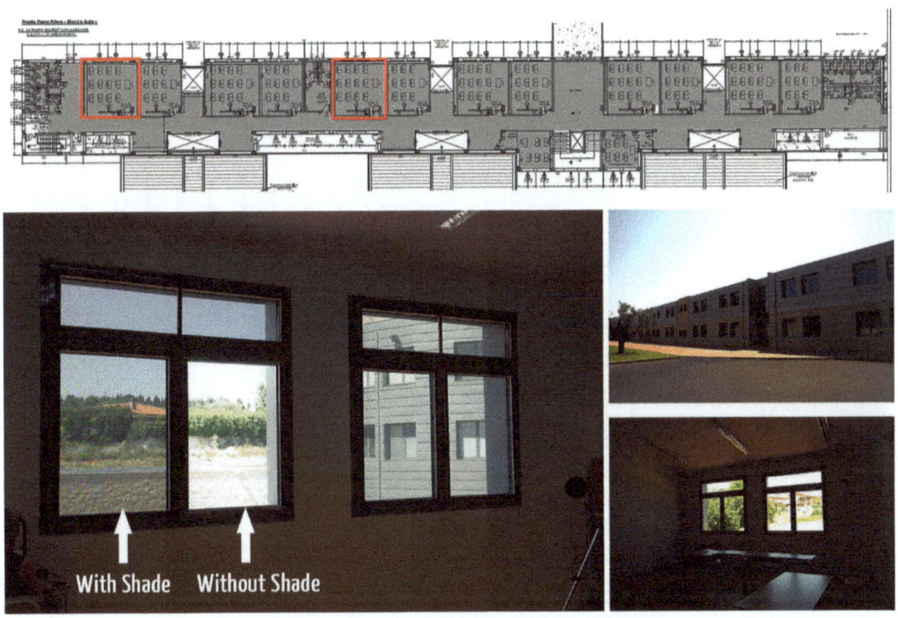

FIG. 1.57 Experimentation of low-e shades on a school building. The case study.

A Window With and Without Shade

The results of the preliminary study (Fig. 1.58) indicate that the internal side of the glass in the case of shaded window is highly overheated and even the internal side of the shade has higher temperatures than those recorded in the inner side of an unshaded glass.

Shaded Versus Unshaded Classroom

The results of the simultaneous comparison between shaded and unshaded classrooms show that the air temperature in the former is higher than in the latter one (Fig. 1.59). This system, which is very useful for winter energy saving, should be designed by providing the opening capability or should be installed on the outer side of the glass (see Data Sheet 1.N).

1.N DATA SHEET: PANELS VERSUS LOUVERS

An experimentation was carried out on a university building by simultaneously measuring the performance of unshaded glass in comparison with a glass with three different shadings: sliding metal panels, aluminum lovers, and external low-e shade (Fig. 1.60).

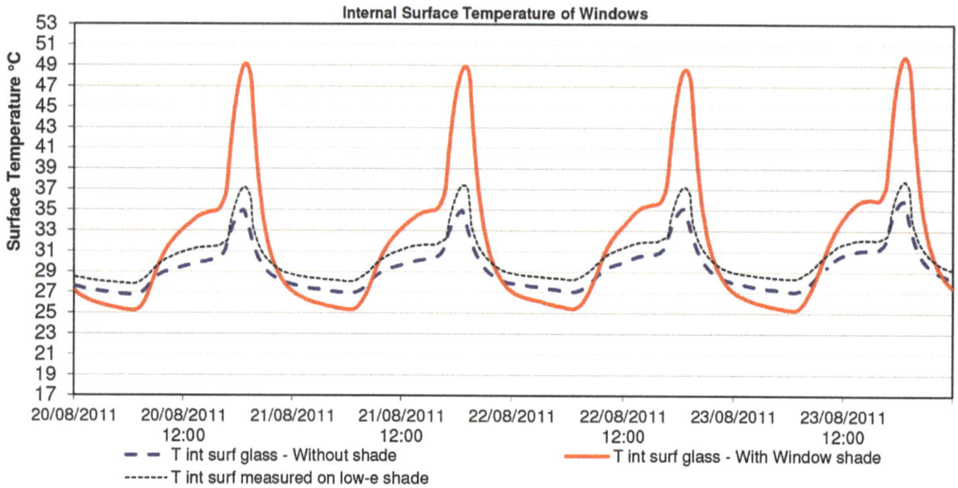

FIG. 1.58 Internal surface temperatures measured on the glasses and on the shade in two adjacent windows, one shaded and one unshaded.

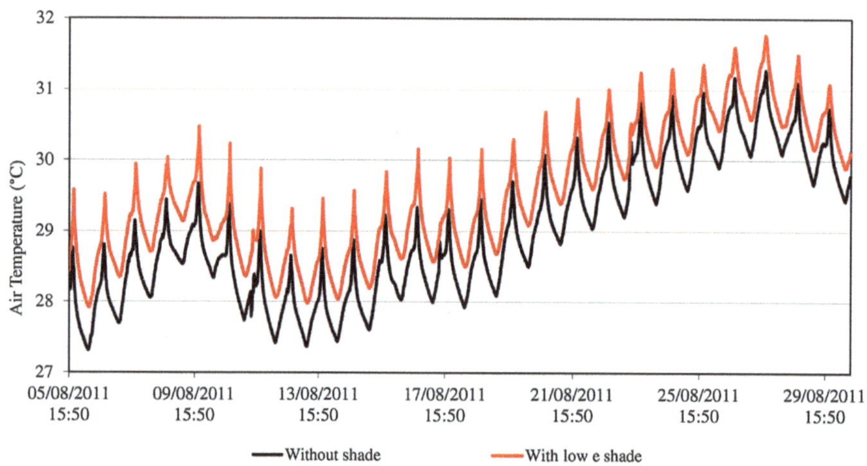

FIG. 1.59 Air temperatures in two adjacent classrooms, with and without internal shade.

The detailed monitoring plan is reported in Section 4.3.1.

The experimentation demonstrated that the use of sliding panels, aluminum louvers, or low-e curtains, on the external side of the window, determines a great enhancement of the thermal performance with respect to the unshaded insulated glass unit. Extended results can be found in (Stazi et al., 2014).

The comparison between the surface temperatures of the glass just behind the three different shadings revealed that the sliding panels and the external low-e shade lower the temperature peaks on the out warding side of the glass down to 9°C (Fig. 1.61).

The deepening of 1 day (Fig. 1.62) shows that low-e panel is able to lower the afternoon overheating peaks but maintains comfortable temperature in working hours, thus resulting even better than the sliding panel.

1.O DATA SHEET: WOODEN LOUVERS VERSUS ALUMINUM LOUVERS

External horizontal louvers have gained the greatest attention in the research field for their high performance in visual light transmission. They are widely used in office buildings as shielding rather than obscuring component.

An experimental and numerical study was carried out on a university building to compare a wooden and an aluminum louvered shading (both characterized by a 45 degrees tilt angle).

The monitoring plan is reported in Section 4.3.1, and the features of the compared shadings are reported in Table 1.26. Extended results can be found in (Stazi et al., 2014).

The comparison among the surface temperatures of the glazed surface behind the two screens at the beginning of the winter period (Fig. 1.63), shows that the glass with wooden persiana reaches temperatures that are slightly higher (about 1–2°C) during the night and afternoon while the overheating peaks during the central hours of the day are slightly reduced. During clear sunniest days, the glass behind wooden system overheats less than the aluminum one: the maximum value at midday is 19°C against 21°C of the glasses screened by the aluminum system, with a reduction of 2°C.

In the interspace between glass and shade, the wooden persiana shows a more conservative behavior, with more flat temperature curve (Fig. 1.64).

This different behavior is due to the specific optical and thermal properties of the shadings (Table 1.27).

The wooden screen presents higher specific heat capacity but far less thermal conductivity than aluminum. The lower conductivity of the wooden screen determines its higher insulating properties and therefore its ability to maintain higher temperatures during the night within the cavity comprised between the glass and the shade. Instead, the higher specific heat capacity of the wooden shade determines its higher capacity to store the heat and release it in a delayed time (during the night). Even if much higher density than wood characterizes aluminum, this property does not influence the storage capacity of the screens as the aluminum slats are very thinner than the wooden ones. Moreover, the two materials present very different emissivity values ($\varepsilon = 0.95$ for wooden screen while $\varepsilon = 0.82$ for anodized black aluminum). Consequently, the wooden system is able to reemit more absorbed energy by radiation, thus lowering the maximum surface peaks.

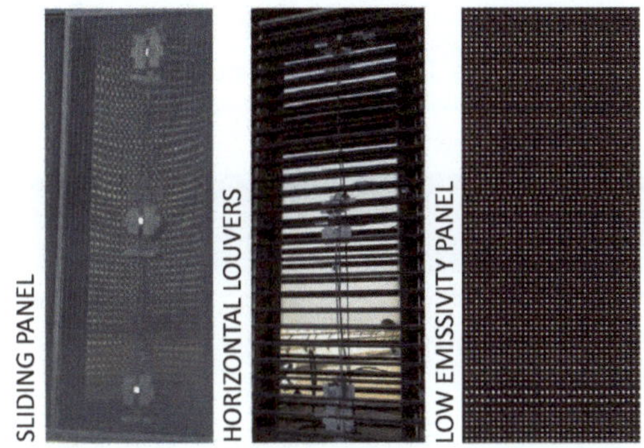

FIG. 1.60 Experimented shades: sliding panel, aluminum louvers, and low emissivity panel.

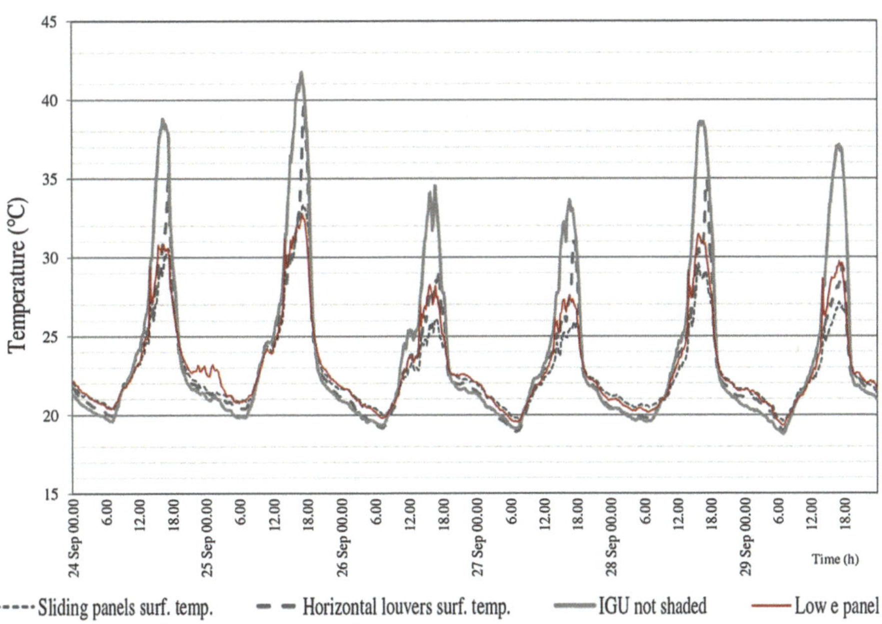

FIG. 1.61 Surface temperatures of the glasses behind the three shadings. September.

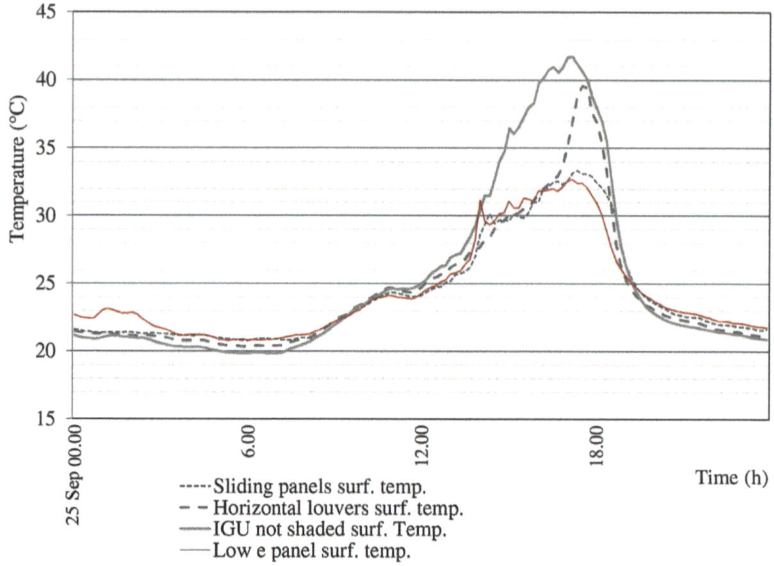

FIG. 1.62 Surface temperatures of the glasses behind the three shadings. Deepening on one day, 25th of September.

TABLE 1.26
Features of the Experimented Louver Shadings.

	Type	Material	Thickness	Width	Angle[a]	Distance[b]
	Alum-45°	Aluminum	2 mm	5 cm	45°	5 cm
	Wood-45°	Wood	8 mm	5 cm	45°	5 cm

[a] Angle formed rotating clockwise from the vertical plane until the slat axis.
[b] Vertical distance between slats.

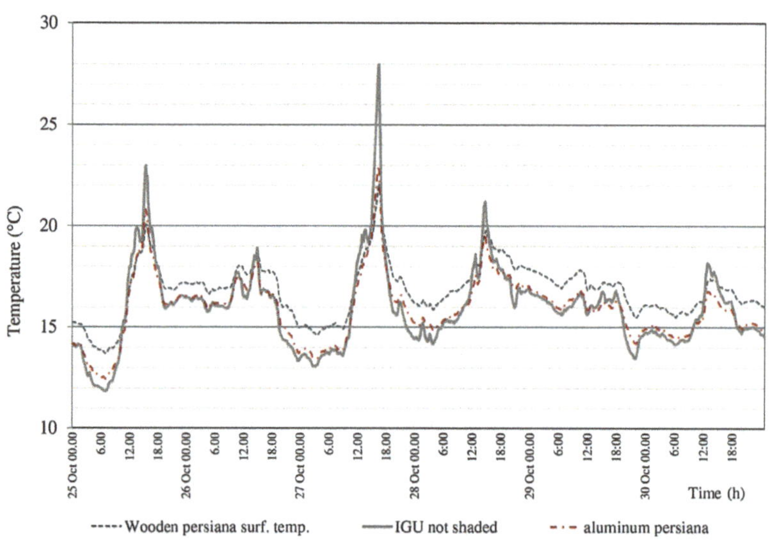

FIG. 1.63 Surface temperatures of the glazed surface behind the two screens at the beginning of the winter period.

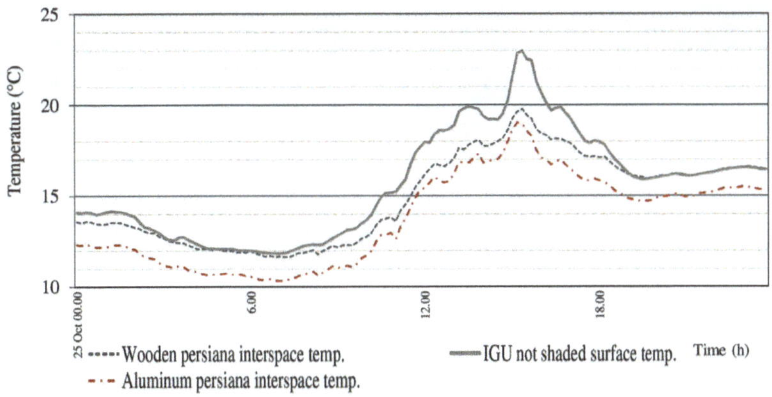

FIG. 1.64 Comparison of the temperatures recorded in the interspaces between glasses and shades on 25th of October.

TABLE 1.27
Optical and Thermal Properties of the Two Shadings.

	Parameter	Aluminum	Wood
$T_{e,B}$ [a]	Solar transmittance (–)	0.015	0.075
$\rho_{e,B}$ [a]	Solar reflectance (–)	0.075	0.375
$\alpha_{e,B}$ [a]	Solar absorptance (–)	0.91	0.55
g_t [a]	Total solar energy transmittance (–)	0.105	0.097
ε	Emissivity (–)	0.82	0.95
λ	Conductivity (W/mK)	237	0.12
Cp	Specific heat capacity (J/kgK)	896	2700
ρ	Density (kg/m^3)	2710	500

[a] According to UNI EN ISO 13363–1:2003.

The wooden screen absorbs the heat during the warm central hours of the day (therefore lessening overheating peaks) and gives it back (in major quantity than the aluminum screens) in less critical hours (with a time shift) thus maintaining higher temperatures at night.

This behavior is beneficial both in winter and in summer periods as demonstrated by the consumption values obtained through the simulations in dynamic state (calibrated with experimental values) reported in Table 1.28. The data show the better performance of the wooden shade (named W-45°). Considering the natural lighting levels, the two solutions are very similar.

1.P DATA SHEET: SLAT INCLINATION AND WIDTH

An experimental comparison was carried out on aluminum louvered shades with different slat angles, namely horizontal (an angle of 90 degrees with respect to the vertical axis) and 60 degrees clockwise with respect to the vertical. Other results can be found in (Stazi et al., 2014).

The experimental measures (Fig. 1.65) demonstrate that the inclination of the louvers allows reducing the daily peaks recorded in the glass behind the shade down to 5°C.

Simulations in dynamic state (Energy Plus software calibrated with experimental values) allowed exploring other slat angles (Table 1.29).

Moreover, by varying the vertical distance between slats and their depth (maintaining a ratio around 1), it was possible to explore narrow and long slat geometries.

The simulations revealed (Table 1.30) that low slat angles, with louvres almost close to the vertical line (i.e., AS-20°), determine, as expected, high heating demand and low cooling consumptions. Inversely, high

TABLE 1.28
Energy Consumptions (for Heating and Artificial Lighting), Comfort Levels, Daylighting Factor (DF), and Light Uniformity (U Ratio) of the Shades.

Shading Device	Heating (kWh/m^2)	Summer Overheating (hours/year)	Art. Light (kWh/m^2)	DF Med (%)	U Ratio (%)
Unshaded	72	545	3	2.5	0.12
AL-45°	72	496	10	2.4	0.1
W-45°	71	310	10	2.4	0.1

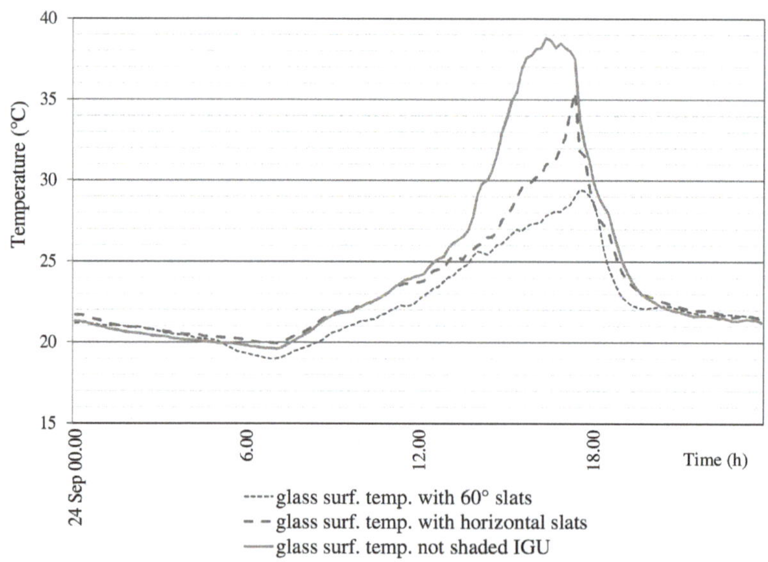

FIG. 1.65 Surface temperature in the glass behind the shade for different slat inclinations.

TABLE 1.29
Explored Configurations Through Numerical Simulations.

Shading Device	Material	Slats angle[a] (°)	Distance Between Slats (cm)	Slats Depth (cm)
AS-20°	Aluminum	20	8.5	8.5
AS-45°	Aluminum	45	8.5	8.5
AS-60°	Aluminum	60	8.5	8.5
AS-90° (horizon.)	Aluminum	90	8.5	8.5
As-narrow slats	Aluminum	45	5	5
As-medium slats	Aluminum	45	8.5	8.5
As-long slats	Aluminum	45	20	15

[a] Angle formed rotating clockwise from the vertical plane until the slat axis.

TABLE 1.30
Energy Consumption and Comfort Levels for Different Slat Configurations.

Shading Device	Heating (kWh/m^2)	Cooling (kWh/m^2)	Discomfort (hours/year)	Art. Light (kWh/m^2)	DF Med (%)	U Ratio (%)
Unshaded	72	56	545	3	2.5	0.12
AS-20°	72	40	543	12	2.4	0.1
AS-60°	71	42	427	9	2.4	0.08
AS-45°	72	41	496	10	2.4	0.12
AS-90°	70	46	413	8	2.4	0.11
AS-narrow	72	42	426	10	2.4	0.09
AS-med	72	41	496	10	2.4	0.1
AS-long	71	43	333	9	2.4	0.11

sun permeability with louvers along the horizontal line (AS-90°) determines low winter consumptions and high summer cooling demand. The natural light is slightly influenced by the slat angles, with artificial lighting consumption increments for nearly closed configurations. The uniformity of light distribution within the room U is enhanced with intermediate positions.

The results regarding the slats elongation indicate that the variation has low incidence on winter consumptions but greatly influences the internal comfort conditions, with overheated time reduced to 333 hours for the long slat configuration.

1.4.2 Dynamic Design

Several authors demonstrated that the adoption of a dynamic solar shading, guided by automatic control of the louvers could maximize the performance with respect to the conflicting aspects of cooling, heating, and artificial lighting consumptions.

Kinetic shading systems move in response to mechanical, chemical, or electrical stimuli. They fold, expand, shrink, and transform. The most widely adopted types are rotating and shape morphing shading systems.

Rotating shadings

The devices are predominantly either glass louvers, metal louvers, or timber louvers and are conceived to rotate around either horizontal or vertical axes. The carrier systems vary with louver size and span; in many applications, aluminum torsion tubes along the entire length of the louvers have been adopted.

Shape morphing shadings

The most advanced research studies on the field regard shape morphing solar shading (Fiorito et al., 2016).

They have different typologies of movement such as translation, rotation, and scaling, as result of the application of an external force. Recent trends regard the replacement of traditional mechanical systems with new systems that integrate multifunctional and smart sensors-actuators responsible of the mechanism control. In such systems, the smart materials act as combined sensor and actuator, to impress to the building component (for example, an aluminum louvre) a movement as response to a specific stimulus.

Stimulus-responsive materials (SRMs) are the most suitable smart materials for this application thanks to their ability to respond to external stimuli through a change in their own physical or chemical properties. They can be grouped into two categories, namely shape change materials (SCMs) and shape memory materials (SMMs). Being only the latter type able to recovery its original shape, it is particularly suited for dynamic building shadings applications.

Shape memory materials can be grouped based on the materials adopted, in alloys (shape memory alloys—SMAs), polymers (shape memory polymers—SMPs), and hybrids (shape memory hybrids—SMHs). Among the several types within each group, the most interesting applications for shadings regard the thermoresponsive types for the simplicity of achieving temperature stimuli with the thermal effect of incident radiation.

SMAs. The most diffused type is based on NiTi (nitinol) alloy, for its corrosion resistance, ductility, and high deformation recovery. The movement consists in small contraction (up to 10%) and deformation. They are durable systems (able to fulfill up to 200.000 cycles). The initial shape is fixed through a thermal treatment, and then the alloy is cooled to relatively low temperatures and plastically deformed; when the alloy is heated it recovers the initial shape because above a certain temperature it behaves as a superelastic material. The phenomenon can be explained with the hysteretic transformation of the alloy during the cooling phase from austenite (parent phase) to martensite (generated phase). This transformation is reversible: during a subsequent heating, martensite transform itself again to austenite.

SMPs. In this case, the base materials are polymers, constituted by chemically bonded molecular chains of monomers. Once the transition temperature is reached, they are deformed through the application of an external force, thus obtaining the desired temporary shape. The subsequent cooling determines the fixing of this temporary shape. Once the material is heated again above the transition temperature, the specimen releases the acquired stress thus returning in its original configuration.

SMHs. According to (Fiorito et al., 2016), the most promising one is the silicon/wax hybrid, being the wax the material assigned at transition and the silicon matrix the one assigned at keeping the elasticity. After cooling, the wax fixes its shape, while the elastic matrix stores the elastic energy. The subsequent heating determines the releasing of stored energy and the restoring of the initial shape.

1.5 OPEN QUESTIONS AND FUTURE CHALLENGES

The new technologies for window systems, regard all building levels, nanolevel (for example new coatings based on nanoparticles), microlevel (such as suspended films), and macrolevel (namely new frames, performing spacers and shadings). All of them hold great promise for the future but, in general, there are still some challenges at the present.

First, the optimal thermal control of transparent components requires the achievement of higher performance levels than actual.

For heating dominated climates, the new frontier regards the development of market accessible windows, with very high thermal resistance. In the residential sector, the target is to develop R-10 windows (0.1 U–factor) while in the commercial sector, the target is to develop R-7 windows.

To that aim, the most advanced studies bet on the development of better thermally broken frames and in the maximization of the thermal performance of all the windows parts, including spacers.

For typical window frames, the glass R-value drops significantly at the insulating glazing unit edge. The adoption of super-low-conductivity frames with thermal efficient new materials (based on aerogel, composites ...) with selected systems to avoid long-term air infiltration will reduce the demand for ultra-high-performance glass. In this case, the use of multiple spacers could lead to air leakage, and it is so preferable to use a unique superspacer with high thermal efficiency, critical also for condensation control. Moreover, the reduction of the gas thermal conductivity was found to have the largest impact on the U-value. The effect gets less pronounced with an increased number of panes in the IGUs.

For cooling dominated climates, it is very important to reduce further the cooling loads by shifting from static properties to the dynamic control of solar heat gain factor and visible transmittance.

The daylighting should be optimized by extending the depth of natural light incoming to reduce electrical lighting. For this aim, also new emerging photovoltaic techniques can be pursued.

In addition to the thermal performance, aging properties and robustness of glazing systems should be further studied.

On the durability level, improvements are needed for vacuum glazing edge seals to maintain gas retention. Moreover, the durability of soft low-e coatings should be enhanced to maintain performance over the expected lifetime of a window. The thermal stresses to the glazing units due to high cavity temperatures can pose a problem for the robustness and lifetime of such units.

The improvement of market acceptability will also involve higher structural performance (design pressures, deflection limits, torsion, etc.) as well as the enhancement of design flexibility, providing the possibility of integration into different wall facades. The weight and thickness of these windows must be comparable with those of the existing stock to enable retrofits of existing buildings. Cavity can be kept at 8 mm in multilayer IGUs to keep the total thickness of the IGU as thin as possible. Moreover, insulating glazing units with several layers of thin glass and/or polymer-based nonstructural intermediate suspended layers could be adopted.

Another critical issue on new emerging technologies regards the necessity to reduce drastically the installed cost. Indeed, the recent dynamic window systems are too expensive for mass adoption in both commercial and residential sectors. Recently, the most advanced research peaks are focusing on intelligent and cost-effective "fenestration" systems in which new materials enable to merge frames elements in unique components. Although, on one hand, they are a result of the search for simplicity to lower the costs, on the other hand, they require great investments for retooling by manufacturers.

One still open problem regards theoretical issues for quantifying the real performance. For example, the means of controlling the spectral properties to achieve the desired transmissivities are currently under prototyping, introducing the difficulties associated with the increase in scale from very small laboratory samples to large-scale commercial glazing of sufficiently high uniformity.

Moreover, the successful provision of the final performance of one very efficient transparent envelope component could not ignore the human interaction with such device. Indeed, the users intervene on building devices triggered by environmental, time-related, and social stimuli. Among the various building operable devices, the interaction with windows has a central role both because the windows opening/closing is demonstrated to be one of the most recurring adaptive action and for its great impact on thermal comfort and building energy consumptions. The reader could find a detailed recognition of the challenges connected to this aspect in Chapter 3.

In summary, a deep understanding of the possible new glazing/shading technologies and the knowledge on how the occupants interact with them is essential to achieve great energy saving and comfortable living environments. Further improving the low-e surface coatings of panes in an IGU yields little improvement possibilities compared with today's state-of-the-art technologies. Further studies should be carried out on improved solid framing materials with lowered thermal conductivity and weight, on the use of polymer layers instead of multiglazing IGU, on the prevention or slowing the aging process by control of temperature peaks in central layer and sealing durability, on the provision of the interactions between users, environment, and dynamic glazing/shadings.

REFERENCES

Al-Obaidi, K. M., Azzam Ismail, M., Hussein, H., & Abdul Rahman, A. M. (2017). Biomimetic building skins: An adaptive approach. *Renewable and Sustainable Energy Reviews*, 1472–1491. https://doi.org/10.1016/j.rser.2017.05.028.

Ando, E., & Miyazaki, M. (2001). Moisture resistance of the low-emissivity coatings with a layer structure of Al-doped ZnO/Ag/Al-doped ZnO. In *Thin solid films* (pp. 289–293). https://doi.org/10.1016/S0040-6090(01)01045-8.

Arici, M., & Kan, M. (2015). An investigation of flow and conjugate heat transfer in multiple pane windows with respect to gap width, emissivity and gas filling. *Renewable Energy, 75*, 249–256. https://doi.org/10.1016/j.renene.2014.10.004.

Atzeri, A., Cappelletti, F., & Gasparella, A. (2014). Internal versus external shading devices performance in office buildings. *Energy Procedia, 45*, 463–472. https://doi.org/10.1016/j.egypro.2014.01.050. Elsevier B.V.

Barbosa, S., & Ip, K. (2014). Perspectives of double skin façades for naturally ventilated buildings: A review. *Renewable and Sustainable Energy Reviews*. https://doi.org/10.1016/j.rser.2014.07.192.

Bessoudo, M., Tzempelikos, A., Athienitis, A. K., & Zmeureanu, R. (2010). Indoor thermal environmental conditions near glazed facades with shading devices e Part I: Experiments and building thermal model. *Building and Environment, 45*(11), 2506–2516. https://doi.org/10.1016/j.buildenv.2010.05.013. Elsevier Ltd.

Buratti, C., & Moretti, E. (2012). Glazing systems with silica aerogel for energy savings in buildings. *Applied Energy*. https://doi.org/10.1016/j.apenergy.2012.03.062.

Cannavale A., Fiorito F., Manca M., Tortorici G., Cingolani R. and Gigli G. (2010). Multifunctional bioinspired sol-gel coatings for architectural glasses. *Building and Environment, 45*(5), 1233–1243. https://doi.org/10.1016/j.buildenv.2009.11.010.

Casini, M. (2018). Active dynamic windows for buildings: A review. *Renewable Energy*. https://doi.org/10.1016/j.renene.2017.12.049.

Chaiyapinunt, S., Phueakphongsuriya, B., Mongkornsaksit, K., & Khomporn, N. (2005). Performance rating of glass windows and glass windows with films in aspect of thermal comfort and heat transmission. *Energy and Buildings, 37*(7), 725–738. https://doi.org/10.1016/j.enbuild.2004.10.008.

Chen, C. C., et al. (2008). Self-cleaning characteristics on a thin-film surface with nanotube arrays of anodic titanium oxide. *Applied Physics A: Materials Science & Processing*. https://doi.org/10.1007/s00339-008-4546-7.

Cuce, E. (2018). Accurate and reliable U-value assessment of argon-filled double glazed windows: A numerical and experimental investigation. *Energy and Buildings, 171*, 100–106. https://doi.org/10.1016/j.enbuild.2018.04.036.

Cuce, E., & Riffat, S. B. (2015). A state-of-the-art review on innovative glazing technologies. *Renewable and Sustainable Energy Reviews*, 695–714. https://doi.org/10.1016/j.rser.2014.08.084.

Cuevas, C., Fissore, A., & Fonseca, N. (2010). Natural convection at an indoor glazing surface with different window blinds. *Energy and Buildings, 42*(10), 1685–1691. https://doi.org/10.1016/j.enbuild.2010.05.003. Elsevier B.V.

Elmahdy, A. H. (2006). Assessment of spacer bar design and frame material on the thermal performance of windows. *ASHRAE Transactions NRC Publications Archive* (pp. 1–14). Québec City: Proceedings of ASHRAE Summer meeting. Q.C., June 24-28.

Fiorito, F., Sauchelli, M., Arroyo, D., Pesenti, M., Imperadori, M., Masera, G., & Ranzi, G. (2016). Shape morphing solar shadings: A review. *Renewable and*

Sustainable Energy Reviews, 863–884. https://doi.org/10.1016/j.rser.2015.10.086.

Granqvist, C. G. (2007). Transparent conductors as solar energy materials: A panoramic review. *Solar Energy Materials and Solar Cells*, 1529–1598. https://doi.org/10.1016/j.solmat.2007.04.031.

Gustavsen, A., Grynning, S., Arasteh, D., Jelle, B. P., & Goudey, H. (2011). Key elements of and material performance targets for highly insulating window frames. *Energy and Buildings*. https://doi.org/10.1016/j.enbuild.2011.05.010.

Marini, A., Passoni, C., Riva, P., Negro, P., Romano, E., & Taucer, F. (2014). Technology options for earthquake resistant, eco-efficient buildings in Europe: Research needs. In *JRC Scientific and Policy reports*. Luxembourg: European Commission Joint Research Centre Institute for the Protection and Security of the Citizen, ISBN 978-92-79-35424-3. ISSN.1831 -9424; http://doi.org/10.2788/68902.

Grynning, S., Jelle, B. P., Gustavsen, A., Gao, T., & Time, B. (2016). Multilayer glazing technologies: Key performance parameters and future perspectives. In P. K. Heiselberg (Ed.), *CLIMA 2016 – proceedings of the 12th REHVA World Congress* (Vol. 2). Vol. 2.

Jelle, B. P., Hynd, A., Gustavsen, A., Arasteh, D., Goudey, H., & Hart, R. (2012). Fenestration of today and tomorrow: A state-of-the-art review and future research opportunities. *Solar Energy Materials and Solar Cells*, 1–28. https://doi.org/10.1016/j.solmat.2011.08.010.

Jelle, B. P., Kalnæs, S. E., & Gao, T. (2015). Low-emissivity materials for building applications: A state-of-the-art review and future research perspectives. *Energy and Buildings*, 329–356. https://doi.org/10.1016/j.enbuild.2015.03.024.

Kamalisarvestani, M., Saidur, R., Mekhilef, S., & Javadi, F. S. (2013). Performance, materials and coating technologies of thermochromic thin films on smart windows. *Renewable and Sustainable Energy Reviews*. https://doi.org/10.1016/j.rser.2013.05.038.

Karlsson, J., & Roos, A. (2001). Annual energy window performance vs. glazing thermal emittance – the relevance of very low emittance values. In *Thin solid films* (pp. 345–348). https://doi.org/10.1016/S0040-6090(01)01055-0.

Li, Y., Darkwa, J., & Kokogiannakis, G. (2017). Heat transfer analysis of an integrated double skin façade and phase change material blind system. *Building and Environment*. https://doi.org/10.1016/j.buildenv.2017.08.034.

Ma, Z., Cooper, P., Daly, D., & Ledo, L. (2012). Existing building retrofits: Methodology and state-of-the-art. *Energy and Buildings*, 889–902. https://doi.org/10.1016/j.enbuild.2012.08.018.

Manz, H., Brunner, S., & Wullschleger, L. (2006). Triple vacuum glazing: Heat transfer and basic mechanical design constraints. *Solar Energy*. https://doi.org/10.1016/j.solener.2005.11.003.

Maroy, K., Carbonez, K., Steeman, M., & Van Den Bossche, N. (2017). Assessing the thermal performance of insulating glass units with infrared thermography: Potential and limitations. *Energy and Buildings*. https://doi.org/10.1016/j.enbuild.2016.10.054.

Misiopecki, C., Bouquin, M., Gustavsen, A., & Jelle, B. P. (2018). Thermal modeling and investigation of the most energy-efficient window position. *Energy and Buildings, 158*, 1079–1086. https://doi.org/10.1016/j.enbuild.2017.10.021.

Oravec, P. (2015). Insulating windows with integrated frame from composite material. *Procedia Manufacturing, 2*, 348–352. https://doi.org/10.1016/j.promfg.2015.07.061.

Pan, C., Sun, C., Wei, H. M., Han, G. Z., Zhang, J. Z., Fujishima, A., & Gu, Z. Z. (2007). Bio-inspired titanium dioxide film with extremely stable super-amphilicity. *Materials Research Bulletin*. https://doi.org/10.1016/j.materresbull.2006.11.008.

Ragesh, P., Anand Ganesh, V., Nair, S. V., & Nair, A. S. (2014). A review on "self-cleaning and multifunctional materials". *Journal of Materials Chemistry A*. https://doi.org/10.1039/c4ta02542c.

Rezaei, S. D., Shannigrahi, S., & Ramakrishna, S. (2017). A review of conventional, advanced, and smart glazing technologies and materials for improving indoor environment. *Solar Energy Materials and Solar Cells, 159*, 26–51. https://doi.org/10.1016/j.solmat.2016.08.026.

Rosencrantz, T., Bülow-Hübe, H., Karlsson, B., & Roos. (2005). Increased solar energy and daylight utilisation using anti-reflective coatings in energy-efficient windows. *Solar Energy Materials and Solar Cells*, 249–260. https://doi.org/10.1016/j.solmat.2004.12.007.

Serra, V., Zanghirella, F., & Perino, M. (2010). Experimental evaluation of a climate facade: Energy efficiency and thermal comfort performance. *Energy and Buildings, 42*, 50–62. https://doi.org/10.1016/j.enbuild.2009.07.010.

Stazi, F., Marinelli, S., Di Perna, C., & Munafò, P. (2014). Comparison on solar shadings: Monitoring of the thermo-physical behaviour, assessment of the energy saving, thermal comfort, natural lighting and environmental impact. *Solar Energy, 105*. https://doi.org/10.1016/j.solener.2014.04.005.

Stazi, F., Giampaoli, M., Nisi, L., Rossi, M., & Munafò, P. (2016b). Mechanical performance reduction of GFRP specimens with polyester matrix exposed to continuous condensation. *Composites Part B: Engineering, 99*. https://doi.org/10.1016/j.compositesb.2016.05.062.

Stazi, F., Giampaoli, M., Tittarelli, F., DiPerna, C., & Munafò, P. (2016a). Durability of different glass coatings in humid and saline environments, ageing impact on heat-light transmission and thermal comfort. *Building and Environment, 105*. https://doi.org/10.1016/j.buildenv.2016.05.029.

Sung, D. (2016). A new look at building facades as infrastructure. *Engineering, 2*(1), 63–68. https://doi.org/10.1016/J.ENG.2016.01.008.

Van Den Bossche, N., Buffel, L., & Janssens, A. (2015). Thermal optimization of window frames. *Energy Procedia*, 2500–2505. https://doi.org/10.1016/j.egypro.2015.11.251.

Van Den Bossche, N., & Janssens, A. (2016). Airtightness and watertightness of window frames: Comparison of performance and requirements. *Building and Environment, 110*, 129–139. https://doi.org/10.1016/j.buildenv.2016.09.034.

Velikov, K., & Thun, G. (2013). Responsive building envelopes: Characteristics and evolving paradigms. *Building Envelopes, Renewable Energies and Integrated Practice*, 75–92.

Wang, J. (Jialiang), & Shi, D. (2017). Spectral selective and photothermal nano structured thin films for energy efficient windows. *Applied Energy*, 83–96. https://doi.org/10.1016/j.apenergy.2017.10.066.

Xing, Y., Hewitt, N., & Griffiths, P. (2011). Zero carbon buildings refurbishment – a hierarchical pathway. *Renewable and Sustainable Energy Reviews*, *15*(6), 3229–3236. https://doi.org/10.1016/j.rser.2011.04.020.

Ylitalo, H., Ip, K., & Marshall, D. (2006). Thermal perfomance of external roller blinds retrofit for offices en the United Kingdom. In *Proceedings of the 3rd international conference on engineering, project and production management EPPM2012, university of Brighton, UK, 10–11th September, 2012* (5). Available at: http://eprints.brighton.ac.uk/11105/1/EPPM Paper External roller blind_4.pdf.

Zhang, C., Gang, W., Wang, J., Xu, X., & Du, Q. (2018). Experimental investigation and dynamic modeling of a triple-glazed exhaust air window with built-in Venetian blinds in the cooling season. *Applied Thermal Engineering*. https://doi.org/10.1016/j.applthermaleng.2018.05.050.

Degree and Doctoral theses:

Fiorelli, A. (2013). *Ottimizzazione del vetrocamera mediante l'utilizzo di biossido di silicio: Sperimentazione in laboratorio e simulazioni energetiche*, unpublished degree thesis, Supervisors: Munafò P., Stazi F. and Di Perna C.

Illuminati, D. (2013). *Ottimizzazione energetica del vetrocamera mediante l'uso di aerogel all'interno dell'intercapedine*, unpublished degree thesis, Supervisors: Munafò P., Stazi F. and Di Perna C

Ursini Casalena, A. (2012). *Design of a sustainable building enclosure: Experimental and numerical analysis of the solar heat-reflecting shields for the home comfort and energy saving*. Unpublished Doctoral thesis.

Patent

European Patent EP 3 071 775 B1, (2019), Bullettin 2019/14, Invisible Window frame, Inventors: Munafò P., Stazi F., International application n. WO 2015/071882 (2015); Certificate of national (IT) concession No. 0001421037 (2013).

CHAPTER 2

Opaque Envelope

2.1 BACKGROUND: TOWARD A CLIMATE ADAPTIVE FOURTH SKIN

Even the opaque envelope, as the transparent part, is undergoing major changes.

Conventionally, it has been considered a thermal barrier, a static protective enclosure, and a self-centric building capsule insensitive to its surroundings; however, for its position at the boundary between inside and outside, it has missed the opportunity to take advantage by the variability of the external environment.

For many years, technological restrictions have limited the application of such principle but the rapid advances in material sciences have offered the possibility of transforming the external wall into a polyvalent lightweight dynamic membrane composed of multi-functioning materials and interactive components.

These new envelopes, defined in the literature as climate adaptive building shells (Loonen et al., 2013), have been applied with several options in sun shading and glazed surfaces (see Chapter 1), while their application on the opaque wall is more difficult. Moreover, the high insulation levels reached in the last decades have decoupled the internal and external envelope faces. So the external shell would interact with external environment, while the internal surface will exhibit an adaptive/dynamic behavior with respect to the internal environment.

Buildings that typically only controlled the interior comfort of its occupants can now play an active role in controlling the exterior comfort of urban landscape (Sung, 2016).

Indeed the researchers are now facing a new social challenge regarding the mitigation of the urban heat island (UHI) phenomenon. It concerns the higher temperatures in urban areas compared to the surroundings. The main cause is the increased urbanization characterized by elevated density of buildings that create urban canyons, the use of heat-storing materials, and the lack of green spaces and water for evapotranspiration.

The intensity of the UHI phenomenon, measured by the maximum temperature difference between the urban and rural environments, could reach values of almost 10°C–15°C in summer time in big cities (Karamanis et al., 2012). In recent years, this phenomenon was found to have a major impact not only on large cities but also on mid-sized and small towns as well.

The effect results in an increase in cooling energy demand of buildings, pauperization of different environmental parameters, deterioration of construction materials, high pollutants concentrations, compromised citizens thermal comfort, and even increase in the number of deaths (Smargiassi et al., 2009).

Consequently, materials and components adopted in the external building envelope side are called upon to play a new huge role in the urban thermal balance.

The reduction of urban temperatures will determine energy saving on cooling demand of buildings, improvement of urban building thermal comfort and benefits on human health. Moreover, the lowering of outdoor heat peaks will limit the degradation of construction materials.

The passive cooling methods that advanced researchers are facing in this moment for the external skins involve the removal of heat using three main processes: (1) convection (for example, ventilation within a layered design); (2) radiation reflection or reemission (spectral design); (3) phase change (for example, evaporation followed by desorption) or other dynamic transformations. Such processes are often borrowed from nature.

The internal surfaces are more and more careful of building occupants comfort. Many researchers are working on internal finishing materials that are able to store heat and moisture dynamically and to interact with users. Solutions with enhanced thermal inertia and encapsulating phase change materials are under study.

Regarding the envelope core, emerging solutions consider the development of holistic technologies that include simultaneously different aspects, such as energy saving, structural, and reduced construction/maintenance operations. For example, structural insulated panels, in the form of preassembled sandwiched structures, are gaining considerable interest. Moreover, the researchers are facing the challenge to integrate several

Advanced Building Envelope Components. https://doi.org/10.1016/B978-0-12-816921-6.00002-0

aspects in retrofit interventions also with lightness constraint.

In summary, the new tendencies are driving toward a dynamic wall, mimicking the nature variability and combining several aspects and subcomponents in a unique product, meaning the sustainability issue as a technological simplification that reduces the use of raw materials, the processes involved, and the end-of-life operations.

This chapter reviews advanced and innovative components of the opaque building skins and reports experimental data on representative solutions. It is subdivided into three sections regarding the external side (Section 2.2), the internal side (Section 2.3), and the central core (Section 2.4) of the building external envelope.

Section 2.5 underlines the major issues and new challenges.

2.2 EXTERNAL SKIN

This section introduces alternative technologies for the external side of sustainable opaque skins. It addresses four main design types: layered, spectral, dynamic, and multifunctional design.

- *Layered design* regards the adoption of multiple membranes to enhance the wall adaptability and responsiveness to the external environment. Advanced techniques include ventilated layers, lightweight composite materials, and structural adhesives.
- *Spectral design* focuses on radiative heat transfers; it generally deals with tint or coatings able to operate a spectral selection of the incident radiation.
- *Functional design* allows achieving optimal behavior on different aspects, such as maintenance, visual comfort, and energy saving with engineered coatings or multifunctional boxes.
- *Dynamic design* concerns the newest emerging technologies that can change their properties stimulated by environmental changes or external energy sources.

2.2.1 Layered Design

The layered design includes the solutions based on the adoption of an external counter wall (that often exploits the natural ventilation benefices) or additive volumes. The newest solutions are based on lightweight composite materials with structural adhesive connections.

Ventilated skins

Double envelopes have undergone a transformation into interstitial spaces that shelter the external wall. These buffer spaces increase the breadth of the entire building envelope to mitigate the extreme fluctuations of temperature through the control of natural ventilation and solar radiation. Sometimes this environmentally variable interstitial space could be occupied by people and might be a sound design option to improve energy efficiency in the retrofit of existing buildings.

In most applications, the external membranes provide the use of ventilated skins based on solar chimney principle (Fig. 2.1).

In the summertime, the thermal gradient within the bottom and upper part of the air gap activates a buoyancy-driven airflow thus reducing the heat gains toward the indoor. Moreover, the stack effect reduces the heat stored by the external wall, thus contributing to the mitigation of the UHI phenomenon and improving outdoor comfort.

Indeed, the major cause of UHI is that most types of pavement and building claddings are fabricated with low albedo and high thermal retention capacity. Therefore, they store the heat in the day hours and the surfaces remain warm even at night being incapable to release it.

Lowering the external surface temperatures by ventilation would be an effective method to control the urban environment overheating.

In the wintertime, the ventilation gap behaves as a thermal buffer reducing the thermal gradient between inside and outside, hence limiting the heat losses.

The behavior of different ventilated facades is experimentally compared in Data Sheet 2.A.

Polymer composites

Used in the automotive, marine, industrial, and ballistic sectors since several decades, fiber-reinforced polymers have seen increasingly applications in civil engineering buildings since the early 1990s. Nowadays, they are becoming a valuable alternative in various applications, replacing or complementing conventional materials.

The high specific strength, lightweight, low electrical and thermal conductivity, and noncorrodibility are the main advantages compared to steel. For these reasons, fiber-reinforced polymers (FRPs) are increasingly used in building façades, in the form of all-in-one cladding panels, load-bearing pultruded profiles, or adaptive and movable components (Fig. 2.2). Some applications propose the production of entire façade elements in the form of one monolithic FRP composite panel, based on a single production process. In this case, the FRP composite sandwich elements provide both the bearing framework of the building façade as well as the insulation infill contribution. A sealing connection needs also

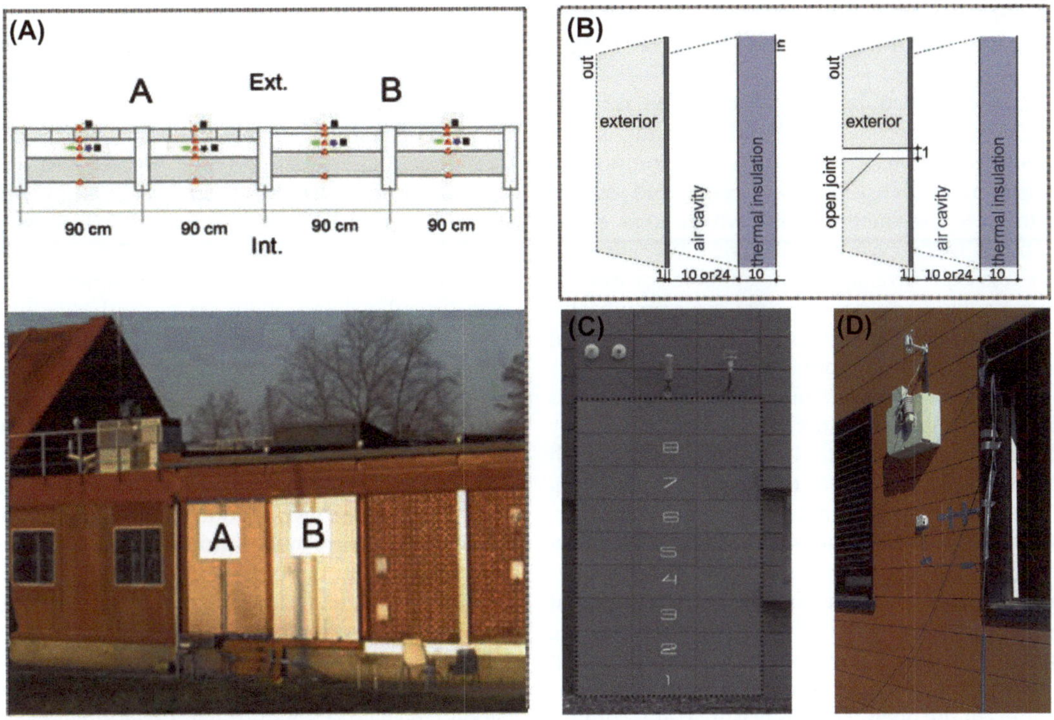

FIG. 2.1 Alternative opaque ventilated facades experimented in the literature: **(A)** alternative claddings from (Langmans et al., 2015); **(B)** variation of joints type, skin materials, and gap width (Marinosci, Semprini, & Morini, 2014); **(C)** and **(D)** experimentations on real case studies, respectively from (Giancola et al., 2012) and (Stazi et al., 2011).

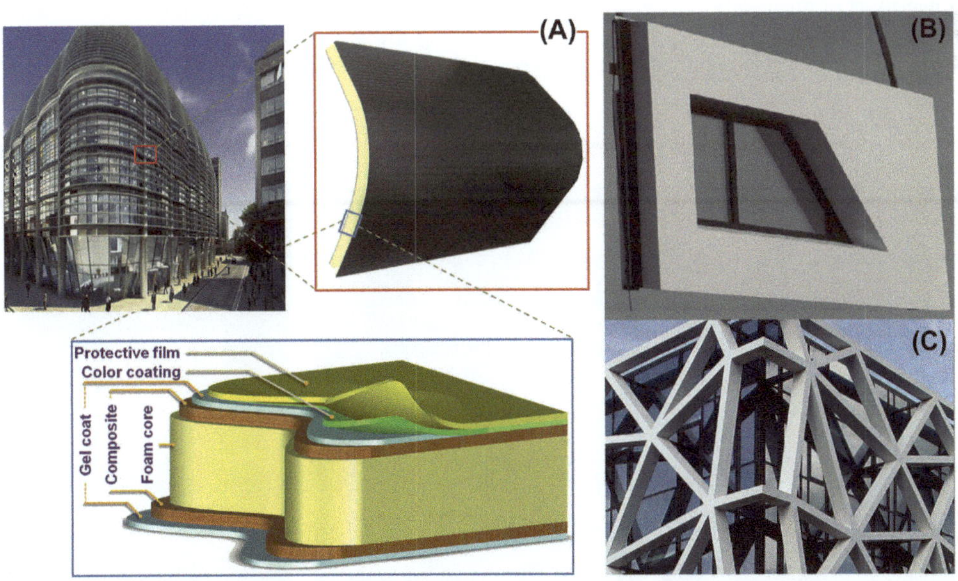

FIG. 2.2 **(A)** Application of composites in building façade and a typical cross-section of a multilayer composite pane; **(B)** composite unitized panels realized in a single production process (Bedon, 2016); composite lattice from https://www.sika.com.

to be applied along the FRP panel edges, so that the single adjacent panels could be properly connected to each other.

Some studies (Bedon, 2016) highlighted that the most suitable composites for buildings and façades are glass fiber-reinforced polymer (GFRP) composites, which uses unsaturated polyester resin, and carbon fiber-reinforced polymer (CFRP), which uses epoxy resin. Among them, even if CFRP is relatively stronger, GFRP is less expensive, and more suitable for exterior applications due to its typically low moisture absorption. Thence, GFRP composites are generally preferred for building envelope components.

The main disadvantage of this material regards instead its orthotropic nature since characterized by high fiber content. Therefore, its stiffness and strength depend on the orientation of the fibers thus requiring the availability of extensive data on mechanical properties. Moreover, a data sheet presented later (Data Sheet 2.B) demonstrates the progressive reduction in profiles performance after exposure to hot wet environment for the increase in stiffness of the matrix. Finally, the presence of fibers determines the unsuitability of bolted joints between members that require drilling operations that would cut the fibers. For this reason, GFRP components usually rely on adhesive jointing.

Structural connections: adhesives

Adhesive connections represent the most efficient jointing method, leading to a uniform load transfer. Their structural integrity is of great importance for the safety of the entire system.

However, some durability issues are still unresolved. Indeed adhesive connections are strongly influenced by environmental conditions, which can reduce their mechanical properties thus compromising the stability of the adherents.

Several authors in the literature investigated their behavior under different types of accelerated environmental conditions. The researchers agreed on the fact that the effect of UV radiation on GFRP profiles is mainly confined to few microns of the surface, leading to very small changes in the mechanical properties. On the other hand, there are dissenting opinions on the effects of the exposure under "hot–wet" environments. Some authors stressed that this type of composites have been successfully used under corrosive environments; others studies, instead, demonstrated that this type of application is critical.

The Data Sheet 2.C, regarding the durability of alternative adhesives, gives a contribution on this debate.

There are three major groups of adhesives used in composites and/or metal structural bonding: *polyurethanes, epoxies, and acrylics*. No one type is suitable in all applications, depending on adherent types, bonding shape, boundary conditions, and forces to be applied to the joint during its service life. The adhesive may need good lap-shear strength or good peel strength or both. Lap-shear strength refers to the ability to maintain adhesion when force is applied parallel to the bond line, while peel strength describes the ability to adhere when pulling force is applied perpendicular to the bond line. Ductile adhesives characterized by high elongation to failure (Fig. 2.3), absorb shock loads, shielding the adjacent structure from fatigue. All adhesives will harden in cold temperatures and soften in hot

FIG. 2.3 Structural adhesives of different types during the tensile tests.

temperatures, so they must be selected to have adequate ductility throughout its in-service temperature range.

The glass transition temperature is one of the most important properties of polymeric adhesives as it is the temperature region where the polymer changes from hard, glassy material to a soft, rubbery material. The exposure under high temperatures or UV radiation, in the range of 90–400 nm, usually causes the exceeding of this critical value, leading to softening, polymers bond dissociation, and decreased elastic modulus of adhesives. Therefore, the adhesive joint stiffness strongly decreases.

As the polymers have a higher heat capacity above the glass transition temperature than they do below it, thermal analyses through the heating/cooling of the specimen is used to measure this specific temperature value.

For adhesive comparisons of this work (Data Sheet 2.C), differential scanning calorimeter (DSC) plots were obtained by heating samples from 25°C to 120°C at a rate of 20°C/minute under inert gas flow of 100 mL/minute. The Tg was determined (according to procedure 10.2.9 of ASTM D 3418) as the midpoint temperature between the extrapolated onset temperature (Teig) and the extrapolated end temperature (Tefg).

The graph in Fig. 2.4 reports the Tg assignment of one sample. It is a typical plot regarding the heat absorbed by a polymer against the temperature obtained by the DSC test. After a certain temperature, the plot shifts suddenly (negative flux has a sudden increase), meaning an increase in the heat capacity of the polymer. This happens because the polymer has just gone through the glass transition. This change takes place over a temperature range.

2.A DATA SHEET: COMPARISON OF THREE VENTILATED SKINS

This data sheet aims at demonstrating that the correct choice of the external envelope layers and the adoption of a ventilated gap are able to reduce the external temperature peaks down to 10°C with respect to a traditional solution with External thermal insulation composite system (ETICS). The optimal solution has also the greatest benefices for indoor conditions.

It reports the results of a yearly experimental campaign aimed at simultaneously comparing ventilated facades with different external mass. Other results can be found in Stazi et al., 2018b.

The external west-facing wall of the mock-up presented in Section 4.4 was used to realize three real-scale prototypes:

- The first configuration (L) is a lightweight typology and has no mass. It was conceived with an aerated gap enclosed by a white plastered OSB (oriented strand board) panel, and it was used as reference for comparisons.
- The second configuration (IM) has an internal mass, with the massive layer (in hollow bricks) adjacent to the insulation material and the external gap enclosed by the same white plastered OSB panel.

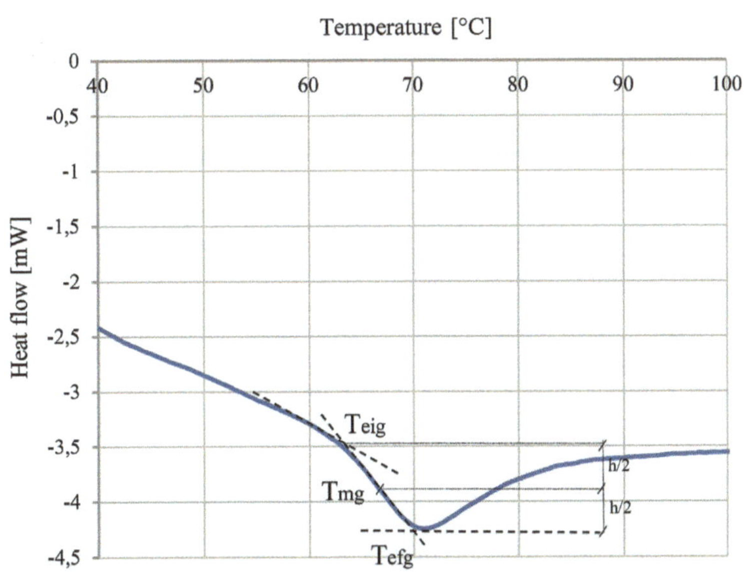

FIG. 2.4 Extrapolated onset temperature (Teig), extrapolated end temperature (Tefg), and midpoint temperature in the middle of the incline (Tmg) from the DSC test. (From Stazi, Giampaoli, et al. (2015).)

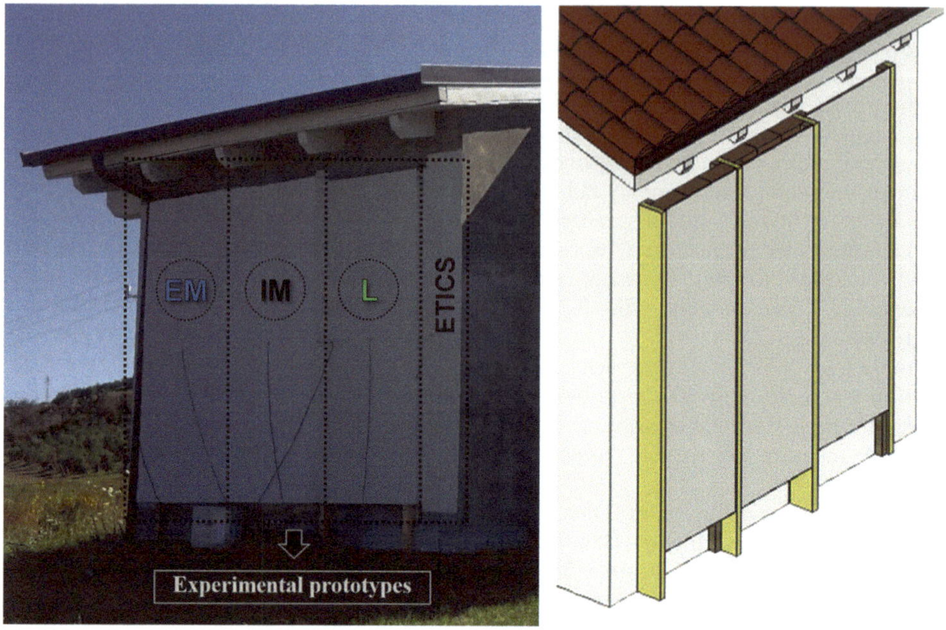

FIG. 2.5 Image of the three experimented prototypes and axonometric scheme.

- The third configuration (EM) has an external mass, namely an outer cladding (in hollow bricks) white plastered. The same surface color was intentionally adopted for the three outer skins to prevent the different optical properties from affecting the results.

One portion of the wall has been left without external ventilation channel to make a comparison with and external traditional insulation composite system (ETICS). An image and an axonometric view of the prototypes are reported in Fig. 2.5. The details of adopted layers and the thermal parameters for each wall are reported in Table 2.1.

The values of the thermal transmittance *U* are obtained according to EN ISO 6946, considering a well-ventilated facade by disregarding the layers between the air gap and the external environment, while κ_2 value is calculated according to ISO 13786, considering all the envelope layers and still air in the gap to highlight the increasing level of external inertia. The mass in IM is expected to have a more conservative behavior as, being enclosed in a cavity, it will have lower opportunities with respect to EM to cool itself at night.

External Surface Temperatures

The experimental measures in May (Fig. 2.6) revealed that a properly designed ventilated skin is able to reduce the heat peaks in the central hours down to 10°C with respect to a traditional ETIC system, thus contributing to the mitigation of the external environment temperatures.

The best choice among the three solutions is the adoption of an external mass (EM solution). This type of skin, differently from the other walls, maintains morning temperatures lower than the environmental values. This is due to its storing mass effectively cooled at night since exposed outdoor. This gives to that wall an advantage, thence it records the lowest values throughout the day.

Other researchers (Alonso et al., 2017) highlighted the beneficial effect of ventilated solutions with external massive cladding in lowering outdoor surface temperatures in both winter and summer. They demonstrated that such solutions could reduce the external surface peaks down to 30°C in winter and about 10°C in summer with respect to ETICS.

Between the three solutions, EM is preferable throughout the year (Fig. 2.7). In summer, it reduced the values of about 1°C with respect to the L solution. In winter, the benefices of adopting an external sub-ventilated massive layer is even more evident. Indeed, the EM wall reduces the surface temperatures with respect to the lightweight L solution down to 3°C in the central hours; the IM solution has intermediate outcomes.

The plotting of the maximum external surface temperature values for a selected day in summer and in winter (Fig. 2.8) demonstrates that the EM solution guarantees the lowest maximum values especially in the winter season.

TABLE 2.1
Drawings, Layers, and Thermal Parameters of the Three Experimented Facades.

	Lightweight (L)	Internal Mass (IM)	External Mass (EM)
Layer	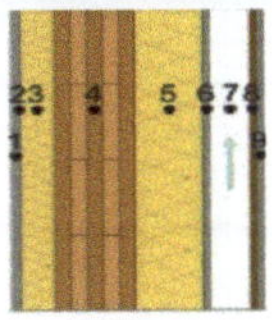	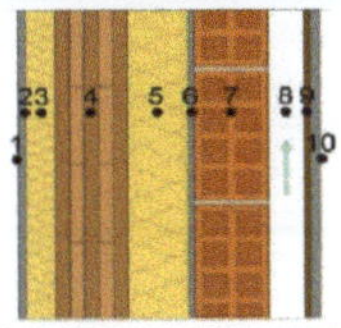	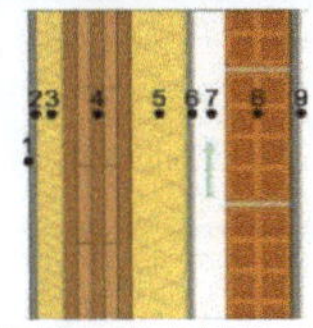
1	Plasterboard	Plasterboard	Plasterboard
2	Vapor barrier	Vapor barrier	Vapor barrier
3	Insulation	Insulation	Internal insulation
4	CLT	CLT	CLT
5	Insulation	Insulation	Insulation
6	Shaving	Shaving	Shaving
7	Air cavity	Hollow bricks	Air cavity
8	OSB panel	Air cavity	Hollow bricks
9	External plaster	OSB panel	External plaster
10	–	External plaster	–
U (W/(m^{2}K))	0.22	0.20	0.22
κ_2 (kJ/(m^{2}K))	28	40	60

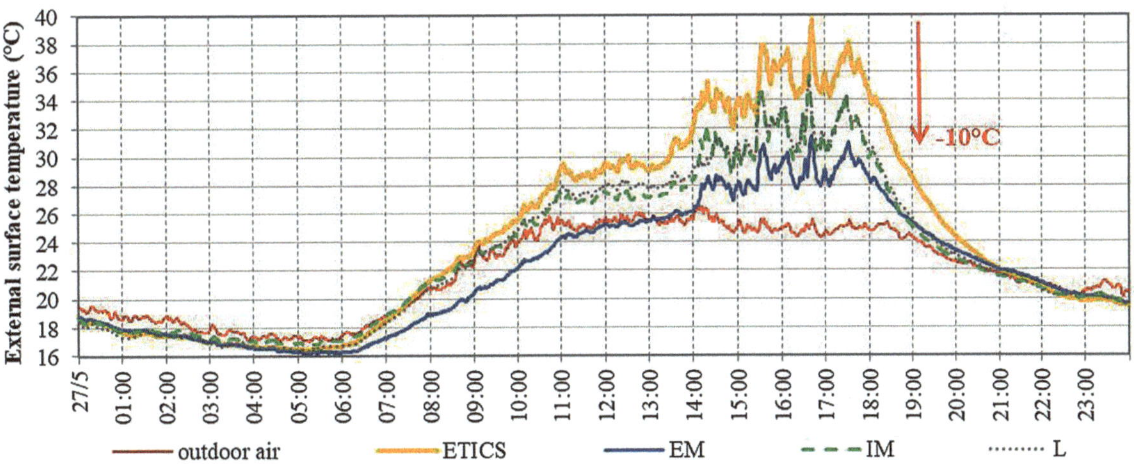

FIG. 2.6 Experimental comparison between three prototypes and the ETIC system (May, 28).

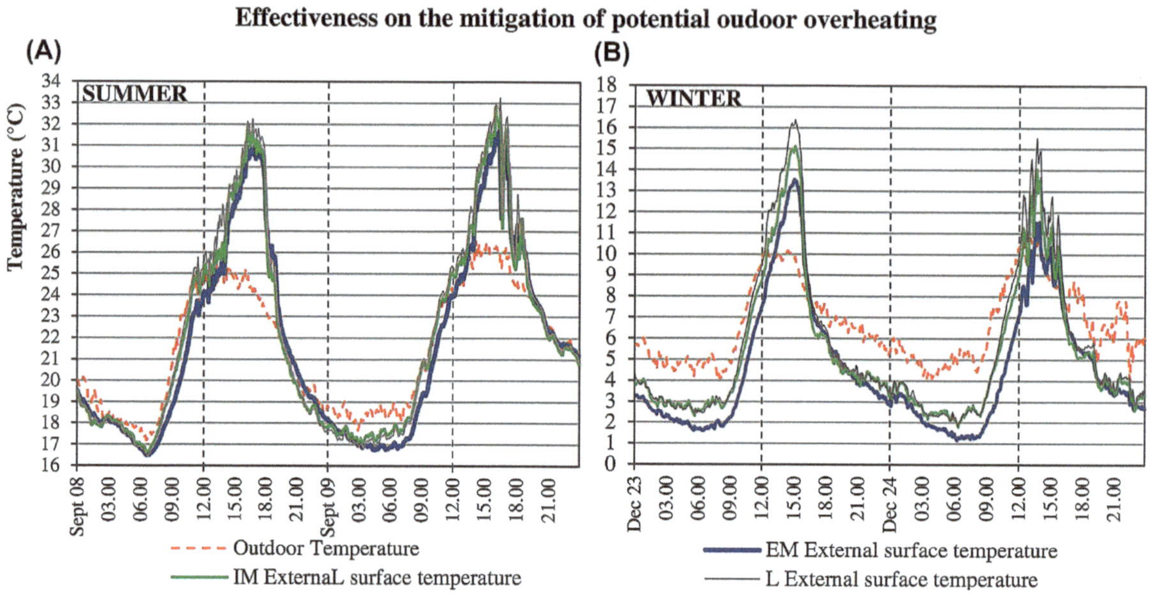

FIG. 2.7 External surface temperatures of the three walls in summer and winter.

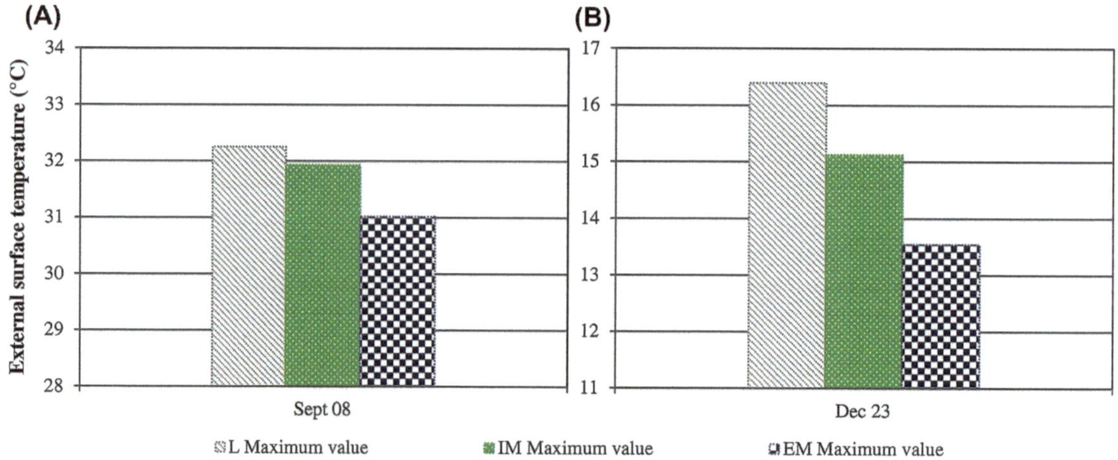

FIG. 2.8 External surface temperature maximum values for the three walls.

Indoor Cooling Performance

For indoor cooling, the study of the heat fluxes in the core of the wall demonstrates adjunctive benefices of the wall with the external mass EM (Fig. 2.9).

Only EM solutions revealed a reduction on incoming heat fluxes. Indeed, for the IM solution the presence of a mass on the internal side of the air gap (instead of externally facing as for EM) determines that the heat stored by the mass, when released, will be directed toward inside causing adjunctive gains. The position of heat flowmeters in the three walls is reported in Fig. 2.10.

As regard the effectiveness of airflow, the comparison of the air speeds within the gap in September (Fig. 2.11) shows that the EM wall has higher performance with a more pronounced chimney effect in the central day hours. The other two walls present lower ventilation regimes in a more distributed time. In summary, the wall with the mass on its outer surface EM can be considered the best one.

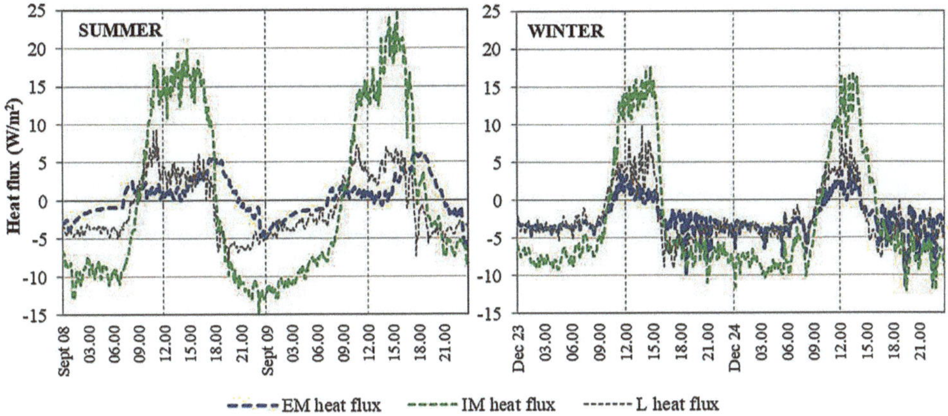

FIG. 2.9 Thermal heat fluxes recorded in the three walls in summer and winter.

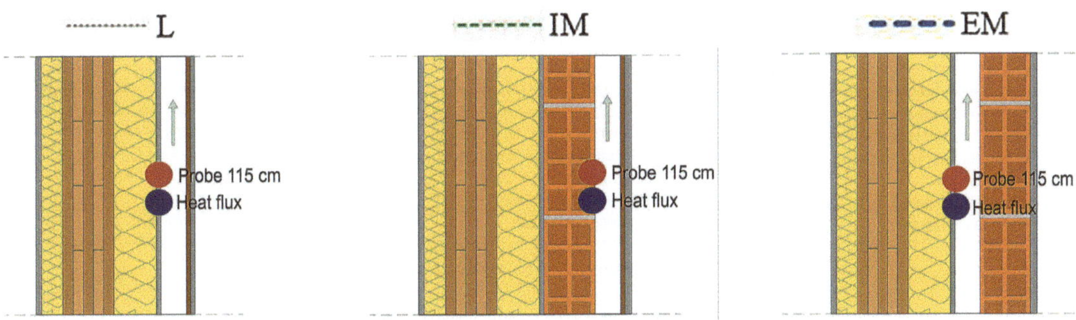

FIG. 2.10 Location of the heat flowmeters in the monitored façades.

2.B DATA SHEET: DURABILITY OF GFRP PROFILES

This data sheet aims at demonstrating that GFRPs undergo a progressive reduction in performance after exposure to hot wet environment for the change in stiffness of the matrix. This is particularly noticeable in all the mechanic tests involving the matrix rather than the fibers (e.g., flexure and torsional tests rather than tensional test along fibers direction).

This data sheet reports the results of an experimental campaign aimed at verifying the effects of hot–wet environments in the durability of glass fiber-reinforced pultruded profiles. Torsional (Fig. 2.12) and flexural tests were selected to highlight the performance reduction after aging. More detailed results can be found in Stazi, Giampaoli, et al. (2016).

Torsional Performance

The results of torsional tests (Fig. 2.13) before (a) and after aging (b) show an almost linear behavior when rotation is between 5 degrees and 10 degrees. Standard deviation (dotted lines) increases with the increasing load, especially in aged profiles.

From torsional tests, the shear modulus of the GFRP profiles could be evaluated using the Bredt-Batho equation for thin-walled tubes. An average value of $G12 = 1.84$ GPa was obtained for the unaged profiles, while a value of $G12 = 1.78$ GPa for the aged ones. Therefore, the aging produced a slight reduction in the shear properties (3%).

The failure modes under torsion (Fig. 2.14) highlight a crack along one of the corner between web and flanges in both aged and unaged samples. However, in the section at the end of profiles, more than one crack occurred in the joining zone, especially in aged profiles.

Flexural Performance

The greatest differences in performances were highlighted during the flexural tests (Fig. 2.15). The diagrams show the load–displacement response of the specimens before (a) and after (b) artificial aging. The mean curves are plotted in the bold format.

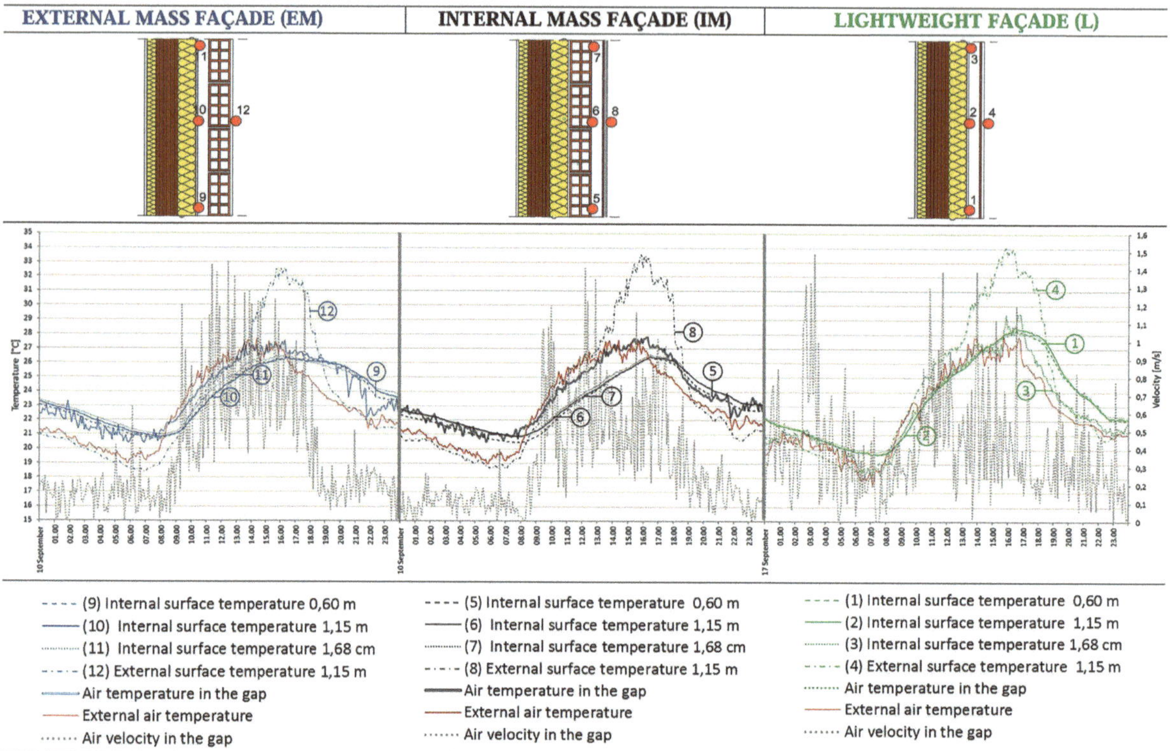

FIG. 2.11 Recorded temperatures and air velocities on one day in September.

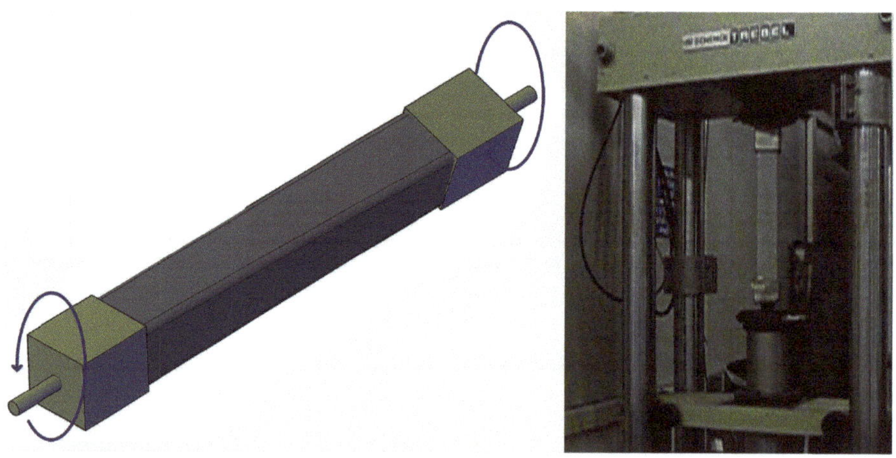

FIG. 2.12 Torsional tests carried put on GFRP pultruded profiles. (From Stazi, Giampaoli, et al. (2016).)

All the samples showed an almost bilinear elastic behavior before failure, with very similar trends, as typically occurring in GFRP under flexural stresses. The failure load for aged profiles was, on average, lower than unaged ones of about 8%.

In bending tests, both fibers and matrix are stressed, and the mechanical properties reduction is mostly attributed to the stiffening of the matrix and the loss of adhesion with fibers, after the aging treatment. Similar results were obtained from other authors

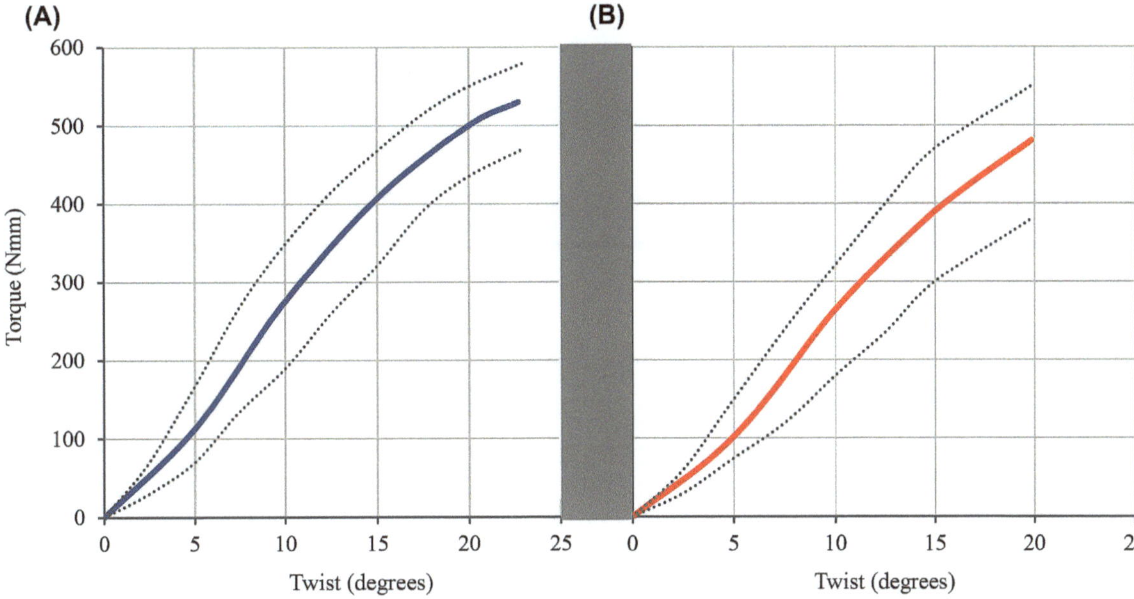

FIG. 2.13 Results of torsional tests: torque against twist: **(A)** unaged condition; **(B)** aged condition. (From Stazi, Giampaoli, et al. (2016).)

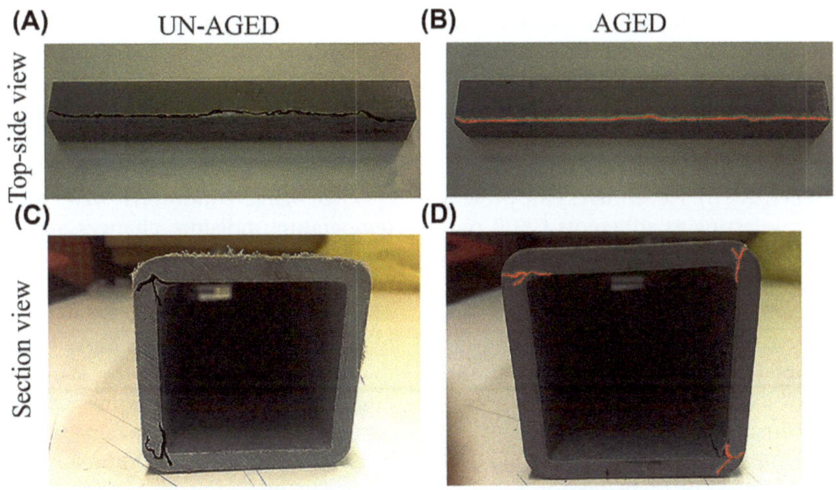

FIG. 2.14 Failure modes under torsion. (From Stazi, Giampaoli, et al. (2016).)

registering a flexural strength decrease of about 10% by exposing polyester profiles to a continuous condensation at 40°C.

2.C DATA SHEET: COMPARISON OF FOUR ADHESIVES TO JOINT GFRPS

This data sheet aims at demonstrating that the choice of the correct adhesive is very important. Indeed the best solution gave double load carrying capacity than the worst one. However, environmental aging has a great effect particularly on this former solution, giving occasion to reflect about what are its most opportune uses.

Validated data about the properties of different adhesives in comparison are very rare (de Castro, 2005). The greatest part of the studies examine the behavior of joints bonded with only one type of adhesive. Furthermore, there are limited data on the aging degradations.

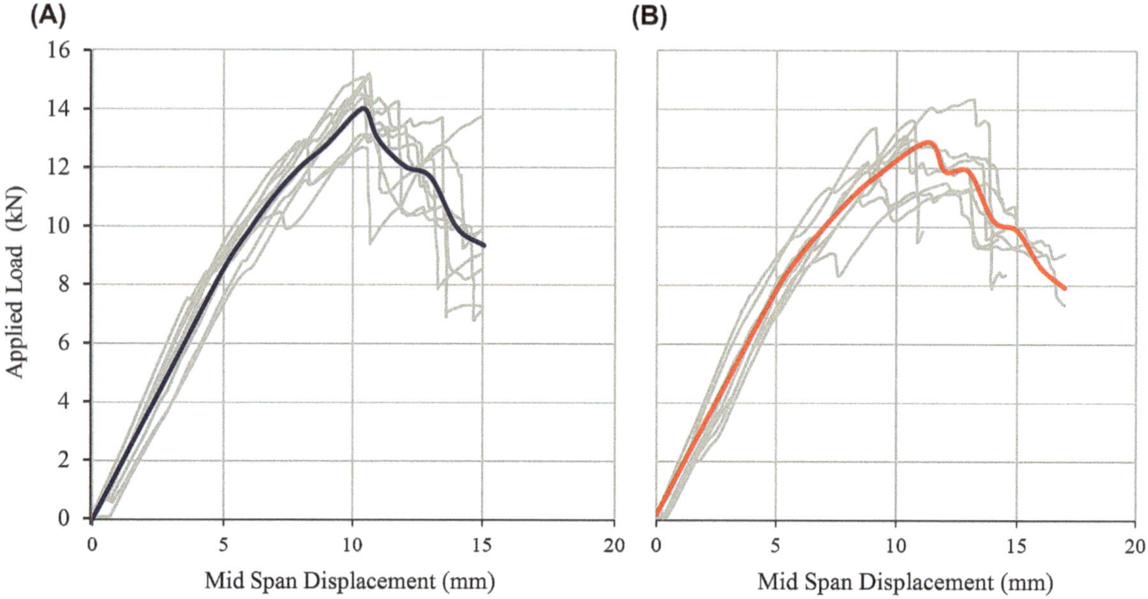

FIG. 2.15 Load–displacement curves for flexural bending tests. (From Stazi, Giampaoli, et al. (2016).)

This data sheet reports the results of an experimental campaign aimed at comparing different structural adhesives for GFRP pultruded joints. A very interesting study has shown that the load–elongation response is mainly influenced by the thermomechanical behavior of the adhesive and much less by that of the adherents (Zhang, Vassilopoulos, & Keller, 2010). Nevertheless the present data are obtained by selecting as adherents GFPR pultruded profiles with vinyl ester resin matrix, more resistant and stiffer than polyester to ensure that the failure mechanism will occur on adhesives rather than on profiles. Four different types of structural adhesives were compared: two epoxy (EPX), one acrylic (ACR), and one methacrylate (MET). More extensive results are reported in (Stazi, Giampaoli, et al., 2015).

To determine the glass transition temperature (T_g) of the adhesives, three samples of each type of adhesive were tested with a DSC. The obtained values (Table 2.2) reveal that the EPX formulations have lower T_g values than the other two adhesives, thus having higher softening risk.

To determine the mechanical properties (E_Y, σ, and ε) of the adhesives, five dumbbell specimens of each type were tested in tension, according to EN ISO 527: 2012.

The shear strength tests were carried out according to ISO 4587:2003 using the single lap joints, while tensile strength tests were carried out according to DIN EN Norm 10002-1 using butt joints (Fig. 2.16 and 2.17). A bonding thickness of 0.3 mm was employed for all the adhesives as recommended by the manufacturers.

A Zwick/Roell Z050 testing machine was used, and an extensometer was applied to measure the strain with a displacement rate of 5 mm/minute.

Two aging conditions were investigated, that is, temperature/moisture changes and UV exposure. The external environment exposure was simulated through cycles using a climatic Chamber. Each cycle consisted of various exposure periods with a combination of high moisture (100% RH) and high temperature (up to 70°C), according to the standard EN ISO 9142: 2003. The UV aging was simulated using eight fluorescent UV lamps. Three specimens were used for each considered aging conditions: no aging (T0), exposure to severe humidity and high and low temperatures (Tc), and exposure to UV radiation (Tuv).

Shear Behavior Before and After Aging

The load–elongation responses of representative unaged and aged specimens, subjected to shear tests (average values), are shown in Fig. 2.18. The curves of the samples aged under severe humidity and high–low temperatures are reported at the center of the graph; the trends of UV aged samples are instead reported on the right.

In unaged condition, all the tested adhesives had almost a similar behavior with elastic nonlinear curves,

TABLE 2.2
Experimented Adhesives.

Adhesive	EPX1	EPX2	ACR	MET
Chemical base	Two-part Epoxy	Two-part Epoxy	Two-part Acrylic	One-part Methacrylate
W_t (min)	>300	120	8–12	15
A_t (°C)	15 to 25	15 to 25	4 to 22	15 to 35
S_t (°C)	−40 + 120	−40 + 80	−40 + 100	−54 + 100
τ^* (MPa)	33.5	15	25	15–30
σ (Mpa)	27.34 ± 0.77	17.11 ± 0.70	11.13 ± 0.50	7.31 ± 0.42
E_Y (MPa)	2966 ± 44	1774 ± 30	648 ± 22	428 ± 62
ε (%)	2.39 ± 0.65	3.81 ± 0.23	7.26 ± 0.70	2.96 ± 0.05
T_g (°C)	61.07 ± 3.34	66.87 ± 0.45	69.37 ± 1.96	70.70 ± 1.41
Use	Structural	Semi-struct.	Structural	Structural

AT, Application Temperature; *EY*, Young Modulus; *ST*, Service Temperature; T_g, Glass Transition Temperature; W_t, Working Time at 22°C; ε, Tensile Strain; σ, Tensile Strength; τ^*, Average Shear Strength on Aluminum/Steel Adherents.
From Stazi, Giampaoli, et al. (2015).

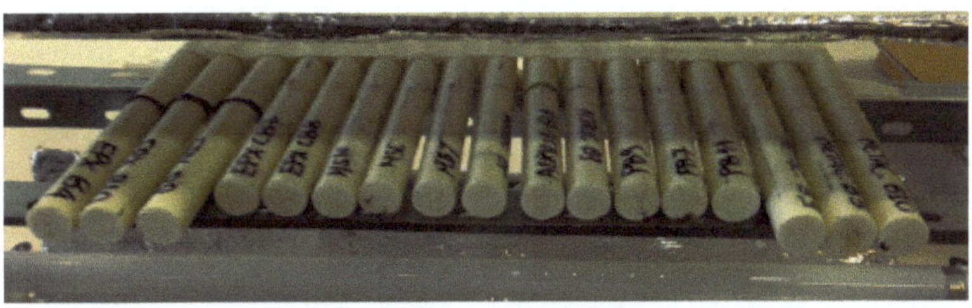

FIG. 2.16 Several butt joints with different adhesives.

(A) (B)

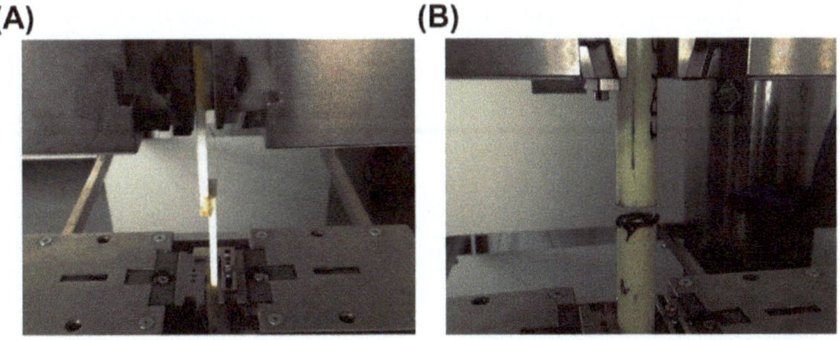

FIG. 2.17 Single lap joints and butt joints, for shear strength and tensile strength tests.

until a brittle rupture. EPX1 showed the greatest performance.

After both the exposures, all the adhesives presented an elastic nonlinear trend, until a brittle rupture. The first epoxy adhesive maintained the highest average strength, while Methacrylate adhesive continued to have the lowest load carrying capacity.

The two artificial aging determined for all the adhesives a large increase of the joints elongations, a slight increase of the average failure loads, and a significant decrease of the joint stiffness (70%–90%). For both environmental exposures, the increase of the load carrying capacity could be ascribed to a further polymerization of adhesives under high or medium–high

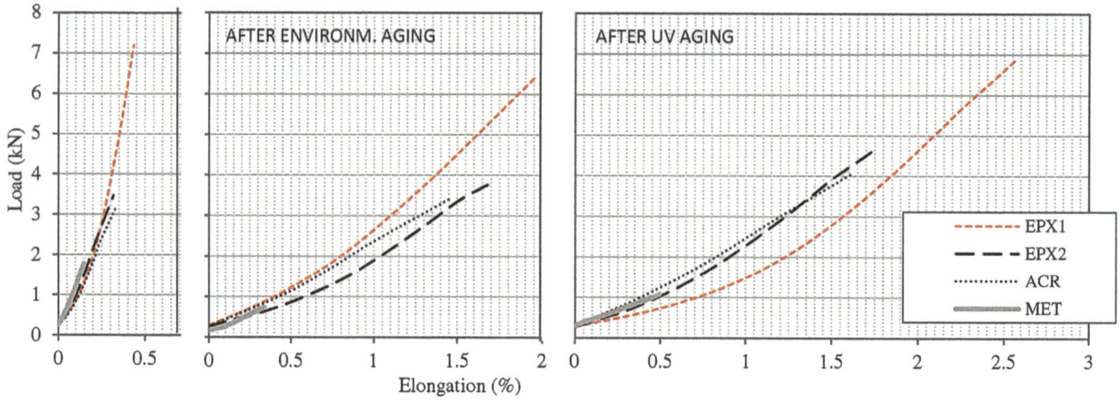

FIG. 2.18 Load–elongation curves in shear before and after the two aging treatments.

temperatures. The joint stiffness decreases during exposition to cycles of different temperatures, and the relative humidity occurred because all the adhesives reached the glass transition temperature. During the UV exposure, the stiffness decrease could be ascribed to the fact that the chemical bonds underwent a dissociation caused by UV wavelengths.

In summary, epoxy adhesives gave the best results in terms of load carrying capacity. However, it is necessary to consider the high deformations that could occur, especially after the UV exposure.

2.2.2 Spectral Design

Spectrally selective materials are designed to have surface temperature values lower than those of typical building materials. Thence they are used to decrease the heat flow entering in the building. Moreover if distributed along the entire building perimeter, they could contribute to lower the air temperature of the urban environment.

Low-e materials

The purpose of low emissivity materials on the external side of the envelope is to reduce the radiative heat transfer incoming from the outside of the building. Indeed these materials are greatly beneficial in warm climate areas where excessive sun radiation is unwanted for larger parts of the year (Jelle, Kalnæs, & Gao, 2015). The use of radiant barriers can be advantageous mainly to attenuate the maximum temperatures reached inside the buildings, as well as the heat flow transmitted through the roofs or walls, in the days of extreme temperatures and with more intense solar radiation.

This technology is based on placing a low-emittance material, facing one or more enclosed air spaces, to reduce the radiation heat exchange between two opposite surfaces. The commonly used solutions regard either the use of reflective insulation layers or paints.

In the case of reflective insulations, the low-e foils (typically being very thin, mirror-like aluminum layers) used as radiant barriers are attached to thermal insulation products that slow down heat transfer also via conduction or convection. The insulation layer is generally fiberglass, foam, or bubble insulation. The ε value goes down to 0.03 in selected products. To perform properly, radiant barriers need to face open space (e.g., air or vacuum) through which there would otherwise be radiation. So they are commonly used in attics or as a vented skin in the exterior of a wall. In this case, a sufficient airspace is required on either side of the foil product (Fig. 2.19). Nonperforated radiant barrier foils behave as vapor barriers while perforated ones allow the vapor to pass through.

Low-e paints are instead paints containing materials that reduce their emissivity value (below 0.25), for example, aluminum powder, silver dust, ceramic beads, or sodium borosilicate microspheres. Even in this case, an airspace of minimum 4–5 cm adjacent to the painted surface is required. Foil products perform better than paints from an energy saving point of view (for the lower reached ε values); however, this latter type can be adopted where the former one is difficult to apply, for example, in cavities of hollow bricks. Experimental studies on such technique are still on course (Fig. 2.20). Some authors (Fantucci & Serra, 2017) (Principi & Fioretti, 2012) demonstrated through experimental comparative measures that a reduction of 18%–20% of the equivalent conductivity can be reached in hollow bricks. Moreover, they measured a reduction of the summer heat loads across roof component of 19%.

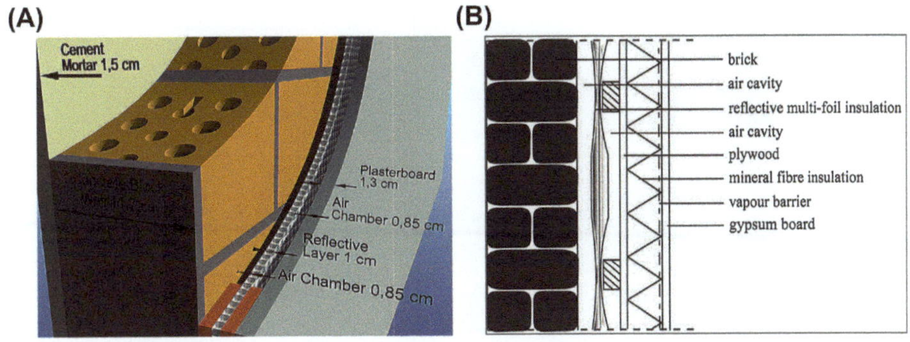

FIG. 2.19 Radiant barrier foils on the external wall (Escudero et al., 2013) and (Jelle et al., 2015).

FIG. 2.20 Experiments on low-e paints on the inner surfaces of hollow bricks (Principi & Fioretti, 2012) and on the inner surface of a roof slat (Fantucci & Serra, 2017).

Cool materials

The increase of the urban albedo is recognized to have a great impact on the long-term global temperature decrease and on the consequent reduction of emitted CO_2 (Akbari, Damon Matthews, & Seto, 2012).

The use of cool materials contributes to the increase of the urban albedo and thus to the enhancement of the climate quality in the cities (Santamouris, Synnefa, & Karlessi, 2011).

Cool materials are characterized by two main properties:

1. High solar reflectivity. It allows the materials to absorb a low quote of solar radiation.
2. High thermal emissivity. High emissivity of the surface of a material means its effectiveness in emitting (i.e., radiating) thermal radiation in the form of long-wave, far-infrared energy depending on its temperature. The faster release of heat impedes the rising of material temperature.

However, highly reflective materials, such as aluminum, have commonly a low emissivity, and dull darker colored (highly absorbing) materials have a high emissivity.

Although reflectivity and emissivity must sum to one at a given wavelength, reflectivity at one set of wavelengths (visible) and emissivity at a different set of wavelengths (thermal) do not necessarily sum to 1. Therefore, it is possible to create visibly dark colored surfaces with low thermal emissivity. White surfaces have instead both requested properties (Fig. 2.21). Moreover, the same above-mentioned research group (Synnefa, Santamouris, & Apostolakis, 2007) satisfactorily experienced the replacement of dark materials with materials characterized by the same color but containing near infrared reflecting pigments.

A parameter than can be used to evaluate a material from this point of view is the SRI, solar reflectance index (Santamouris et al., 2011), as it combines both solar reflectance and infrared emittance in a single value, focusing on the final effect (increase of material temperature) instead of focusing on the cause (material properties). Introduced by ASTM E 1980, it quantifies the behavior of the surface under study with respect to the behavior of a standard black (reflectivity 5%, emittance 90%) and a standard white surface (reflectivity 80%, emittance 90%).

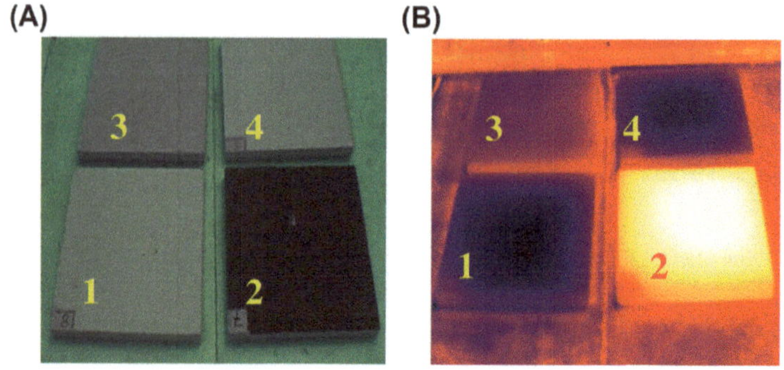

FIG. 2.21 Visible **(A)** and infrared **(B)** image of four concrete tiles painted with cool white materials (1 and 4), a black coating (2) and an unpainted off-white (3) one. The difference in solar reflectance leads to significant differences in surface temperatures. (From Santamouris, Synnefa and Karlessi (2011).)

FIG. 2.22 Scheme of hydrophobic coating functioning principle and inherent self-cleaning nature of lotus leaf (Wang et al., 2011).

$$SRI = \frac{(T_{black} - T_{surface})}{(T_{black} - T_{white})} * 100 \quad (2.1)$$

where T_{black} and T_{white} are the steady state temperatures of the standard black and white material surface, and $T_{surface}$ is the steady state temperature of the material under study.

The higher the solar reflectance index SRI, the higher will be its ability as cooling material.

White coatings, white clay tiles, and white paintings can reach SRI values higher than 100, behaving better than the surface assumed as standard.

2.2.3 Functional Design

Multifunctional solutions are based on multidisciplinary technologies that simultaneously optimize several aspects, such as UV blocking and self-cleaning; thermal insulation and long-term durability; energy efficiency and structural improvement; solar control and air quality.

Self-cleaning coatings

Self-cleaning coatings clean themselves through the action of water.

They can be divided into two categories, hydrophobic and hydrophilic, according to their different behaviors. Hydrophobic surfaces can have also anti-icing, oil repellency, and electrowetting functions, while the hydrophilic coatings are able to break down the dirt chemically. Data Sheet 2.E reports experimental data on both types of coatings.

Hydrophobic coatings. They make the water droplets to slide and roll over the surfaces thus carrying away the dirt. The requirements for this type of coat are very high static water contact angle (>90 degrees) and very low roll-off angle (i.e., low value for the minimum inclination angle necessary for a droplet to roll off the surface). For contact angle approaching more than 150 degrees, the surface is referred to as superhydrophobic (Fig. 2.22).

Hydrophobicity is enhanced by the roughness factor, a feature that can be controlled by the use of nanoparticles. This property is owned in nature by lotus leaf surfaces (Ragesh et al., 2014).

The techniques used for producing hydrophobic surfaces are (a) inducing the roughness on low surface energy materials and (b) patterning of a rough surface with low surface energy materials.

Silicon dioxide-based products are hydrophobic while the combination of titanium and silicon dioxide provided coated surfaces with strong hydrophobicity (Andaloro et al., 2017). This condition may be particularly advantageous in significantly polluted environments with low precipitation rates, where the ordinary self-cleaning of facades through rainfall washing can hardly be achieved.

Hydrophilic coatings. Hydrophilic surfaces have a water contact angle <90 degrees and superhydrophilic are those with a contact angle close to 0 degrees.

This property is owned in nature by pitcher plants that lure insects onto the ultra-slippery surface of its leaves, where they slide (Fig. 2.23).

They are typically photocatalytic materials. Among them, titanium dioxide is widely used for its nontoxicity, availability, cost effectiveness, chemical stability, and favorable chemical and physical properties.

Photocatalytically generated holes (h^+) oxidize the lattice oxygen at the surface of the material resulting in oxygen vacancies that can be filled by adsorbed water. This causes the formation of surface hydroxide groups that increase the hydrophilicity of the surface reducing the contact angle down to 0 degrees (Fig. 2.24).

In addition, the metal oxides have the property to chemically breaking down organic pollutants and complex dirt deposits.

The cleaning mechanism occurs in two steps: photocatalysis, in which organic dirt molecules get decomposed in the presence of sunlight and sheeting of water that carries away the dirt molecules.

Titanium dioxide-based sol has hydrophilic characteristics. Although some products working on the principle of hydrophilicity are commercialized, this topic is far from attaining maturity and there are regular publications in this field.

Air filtering and depollution

Promising research innovations regard the adoption of envelopes able to filter, depollute, and eventually preheat the air (Fig. 2.25).

Actually, some researchers are facing the challenge of promoting fluid motions of the air and attracting pollution, even if most of them are only at its design stage.

Breathe brick is an example of air filtering for indoor air quality. This project involves the use of minicyclones inside the exterior wall cavity to remove particulates from the air (Sung, 2016). Although this proposal is to make a system that filters outdoor air for indoor use, it can also be applied to filter outdoor air for outdoor air quality improvement.

FIG. 2.23 Scheme of hydrophilic coating functioning principle and images of hydrophilic attitude exhibited in nature. (Images from Martin, Brown and Bhushan (2017).)

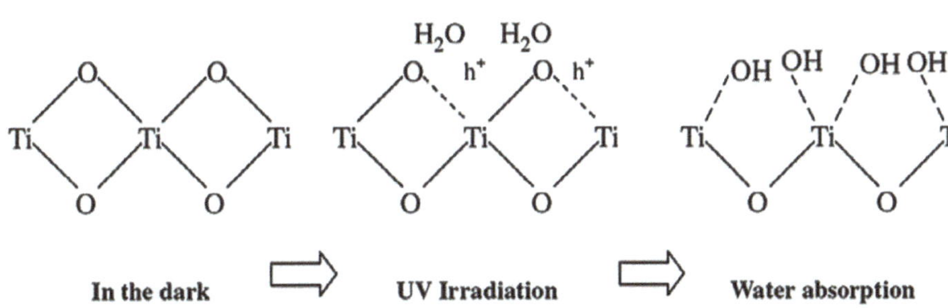

FIG. 2.24 The mechanism for photo-induced hydrophilicity of TiO_2 by the formation of surface hydroxide groups. (From Chen and Poon (2009).)

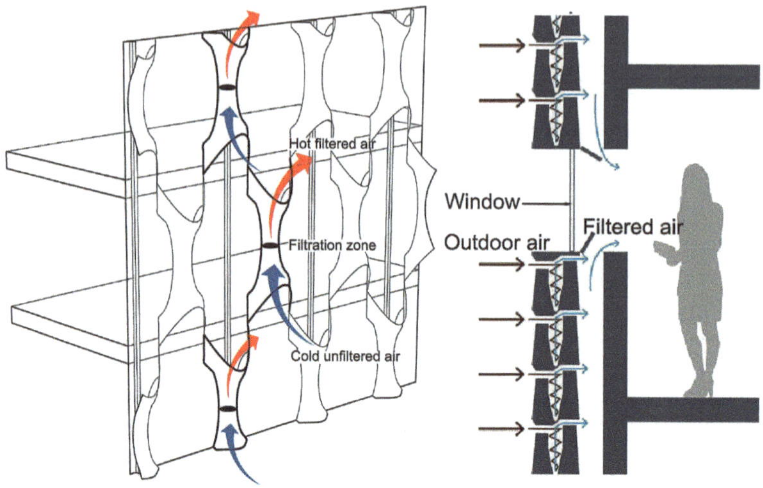

FIG. 2.25 Depolluting walls and breathe-brick wall. (From Sung (2016).)

New holistic skins: from ETIC to BEST systems

ETICS (External Thermal Insulation Composite System) is the most commonly adopted solution for envelope energy retrofit.

However, this solution is not enough to reach the very high energy saving goals that were set (such as zero energy), unless adopting very high thicknesses. It can result in a not cost-effective outcome for a more complex design, construction, maintenance, an adverse net-to-gross floor area, and possible heavier load-bearing constructions.

Advanced studies on the cost-efficacy of retrofitting (Xing, Hewitt, & Griffiths, 2011) suggest bringing to the extreme the thermal insulation properties within lightweight components by adopting emerging insulation materials. A range of very low thermal conductive insulations is actually under study. The studies bet on multilayered insulations, highly performing materials incorporating gas (silica aerogels) or cementitious/polymeric composites with nanofillers or nanofibers in which the nanosize of the particles gives great benefices with respect to their larger scale counterparts (Chen et al., 2013; Shoukry et al., 2016).

However, no response to the structural safety issue is given by adopting such ETICS solutions. In recent years, the new construction codes (e.g., the new Eurocodes) have strengthened the requirements. Most of the existing buildings have already exhausted their design service life (50 years) and require renovation interventions to maintain their service functionality, especially in seismic areas. However, the interventions are often based on a case-by-case approach, without being optimized through transversal cost–benefit evaluations.

The very new approaches toward sustainable seismic retrofits of reinforced concrete framed structures (Negro & Mola, 2017) highlighted that local interventions (beam-to-column joint strengthening) are not cost effective, as they require demolitions of the finishing (which affect up to 70%–75% of the total construction costs) and the temporary downtime of the building. Instead, external global solutions, aimed at complementing the existing structures with new earthquake resistant systems, namely shear walls either over-resistant or dissipative, are very promising from a cost-effectiveness point of view.

Therefore, external insulating over-resistant skins (BEST: building external structural and thermal systems) could be a cost-effective combination of both structural and thermal performance. Integrated interventions will have high positive impacts on the decrease of installation times (for reduced on-site operations) determining overall cost reduction with respect to typical processes commonly split at two different times.

Very recently, some configurations of this type were proposed in the literature, not only for the building sector. They can be grouped into two families:

Engineered multilayered skins. Recently, some researcher proposed new materials with combined structural and thermal performance, mainly for earthquake and fire-combined events.

Some examples (Fig. 2.26): foamed insulations with nanosized additives characterized by high mechanical strength (Saha, Kabir, & Jeelani, 2008; Stazi, Tittarelli, et al., 2018); polystyrene panels with transverse structural connectors (Sassu, Stochino, & Mistretta, 2017); thermal insulating GFRP-reinforced jacketing

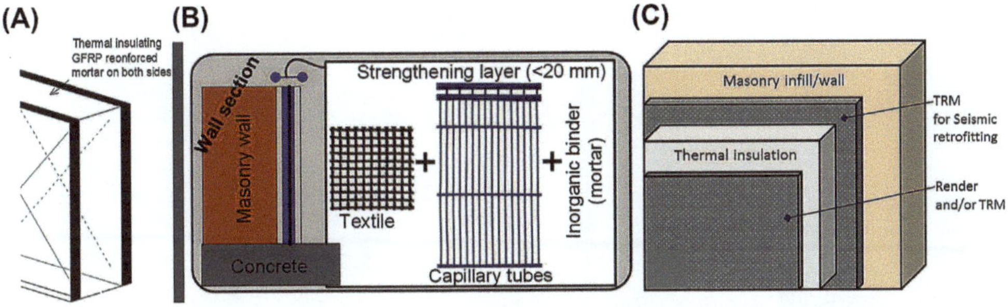

FIG. 2.26 Multilayered and multifunctional skins proposed in the literature: **(A)** Buratti et al. (2016); **(B)** and **(C)** from Bournas (2018).

(Borri et al., 2016); inorganic textile-based composites combined with insulations (Bournas, 2018; Triantafillou et al., 2018).

Moreover some application of multilayered skins intended for both thermal and structural efficacy can be found in aeronautic field for cryogenic systems (Dye, Kopelove, & Mills, 2010). The presence of multiple layers of alternating massive (load bearing) and insulating materials is demonstrated to have high decrementing attitude (Stazi, 2017) thus creating effective thermal barriers.

Data Sheet 2.E reports experimental data on sprayed nanofoams for building thermal insulation aimed at simultaneously enhancing the thermal and mechanical envelope properties.

Energy dissipating double skins. The recent interest toward multidisciplinary retrofit approaches has gathered the attention to structural shear-dissipating double skin facades. Indeed, recent studies of the present author (Stazi, 2017) and other authors demonstrated that using thick insulation on existing massive envelopes could introduce overheating problems, thus suggesting to adopt ventilated air layers with external protective rain screening claddings. Such opaque double skins can be combined conveniently with a shear-dissipating structure.

Some examples of this type were proposed by our (Serpilli et al., 2018) (Fig. 2.27) and other research groups (Dalla Mora et al., 2015; Feroldi et al., 2013).

Multifunctional modules

A large number of innovative solutions for deep renovation have become available from recent EU-funded research projects (Fig. 2.28).

Building additions (ABRACADABRA project), interactive new boxes on the balconies (SELFIE project), and plug and play new façade modules (P2 ENDURE project) are only some examples of parts that can be integrated into the renovation of an existing building.

They usually consist of both opaque and transparent parts. In some cases, they are reduced to window boxes and placed in correspondence to the balconies. Each module can be designed with a specific task or a combination of different functions, for example, purification of incoming air (as a filter), production of renewable energy (using photovoltaic panels), smart control of energy flows based on external conditions, and enhancement of external environment quality. Often, they use smart materials, as phase change materials and nanocomposites (i.e., photocatalytic paints) for the external surfaces.

They behave differently according to the seasons and depending on the incident solar radiation, providing if necessary forced ventilation, changing the storing ability through variable thermal inertia, operating a variable disconnection between indoor and outdoor, and preheating the eventually supplied ventilation air.

Some author also proposed prefabricated algae bioreactor panels, designed to absorb the solar radiation and supply bioelectricity.

Data Sheet 2.F reports experimental data on a preassembled multifunctional ventilated module proposed by our research group.

2.D DATA SHEET: NANOTECHNOLOGY SURFACE TREATMENTS

This data sheet aims at demonstrating that the selection of an appropriate surface treatment for an external plaster could enhance the resistance to water and thus the surface durability.

It reports the results of an experimental campaign aimed at simultaneously comparing different surface treatments for the external plaster. More results are reported in (Stazi, Nacci, et al., 2016).

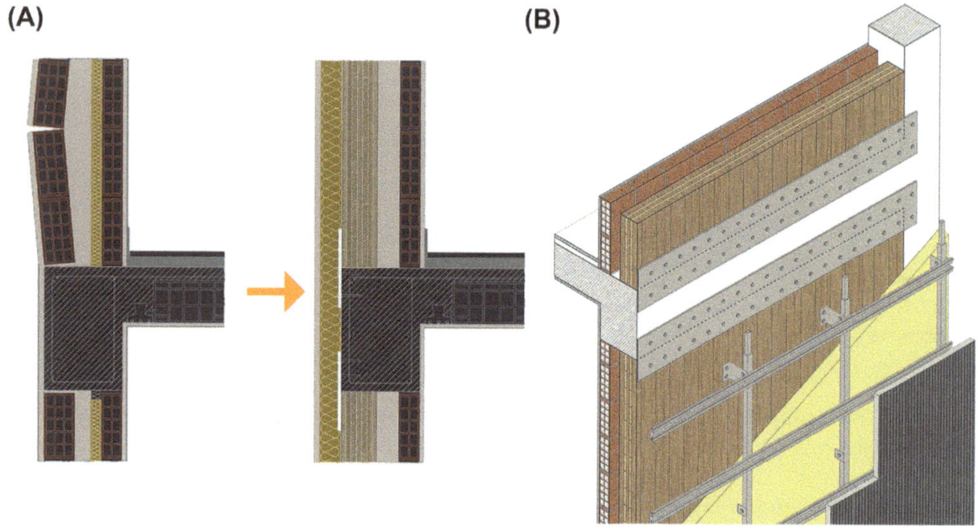

FIG. 2.27 The energy dissipating opaque double skin proposed by our research group by changing the old hollow blocks with cross-laminated timber panels **(A)**. It is simultaneously a seismic resistant structure and an energy retrofit solution, through the insertion of an external ETIC system or an external opaque cladding, thus configuring a ventilated wall **(B)**.

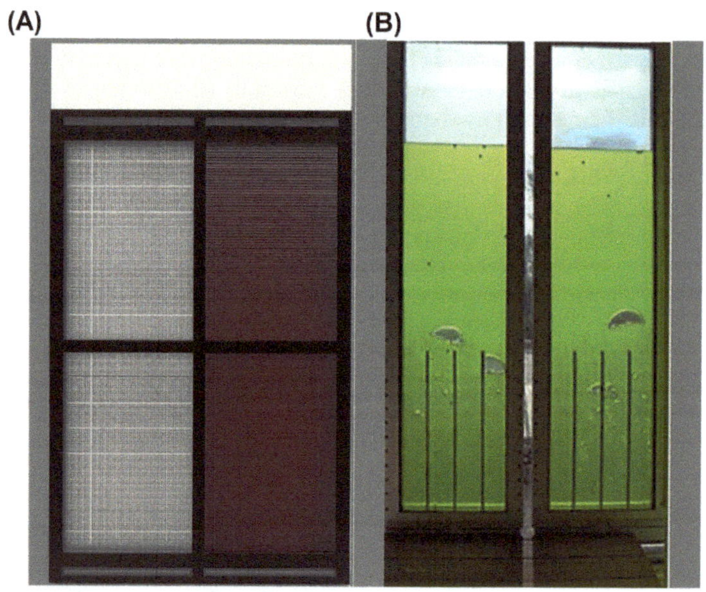

FIG. 2.28 Multifunctional modules proposed in recent studies: **(A)** SELFIE façade solution (Gallo & Romano, 2017); **(B)** Façade with microalgae (Elrayies, 2018).

Four surface treatments were selected among those indicated to protect the surfaces of porous building materials against weathering.

In particular, the compared treatments are the following: hydrophobic silicon nanoparticles, hydrophilic titanium and silica nanoparticles, traditional silane–siloxane treatment, and natural beeswax treatment. The treatment with titanium and silica nanoparticles is to make the surfaces self-cleaning while the other treatments are to make the surfaces water repellent.

The water resistance properties were evaluated through the following tests: wettability, water absorption, and erosion resistance. Moreover, the aesthetic appearance after the surface treatments was evaluated through a surface color test.

Wettability

The wettability was studied by qualitatively evaluating the static contact angle (degrees). It is described as the angle at which the liquid–vapor interface of a water drop meets the horizontal surface of the specimen. When the contact angle is greater than 90 degrees, the surface is considered to be hydrophobic otherwise hydrophilic.

The measurement was conducted on pictures taken with a digital camera in four distinct times, respectively, at 5 seconds, 15 seconds, 2 minutes, and 10 minutes from the first contact between the drop and the specimen.

The results (Fig. 2.29 on the right) show that the silane—siloxane treatment and that with silicon nanoparticles are the sole treatments able to make the plaster water repellent, with contact angles around 110 degrees. The other two behaved as hydrophilic.

Water Absorption

The water absorption was determined using the Karsten pipe method according to the Rilem II.4 recommendations. The tests were performed on small samples (250 × 250 mm) of earth-based plasters applied to a wall panel (Fig. 2.30). This method reveals the attitude of the surface toward water and can be considered as an alternative with respect to the previous test of wettability. However, differently by the latter one, it also simulates the pressure of driving rain, which could be able to "break" the water-repellent behavior of a treated material.

The Karsten pipe is a tube graduated to the 10th of a millimeter provided, in the lower end, with an aperture of 500 mm^2 internal area. The vertical pipe was sealed to the sample with mastic and then filled with water. The initial water column, 92 mm high, exerted a pressure of about 900 Pa in the contact area between the pipe and the sample surface.

The readings of the reduction in the water level were taken every minute for 15 minutes, and then the water absorption rate (C10 minutes, $g/m^2/s$) was calculated using the following formula:

$$C_{10min} = (m_{15min} - m_{5min})/(A \times t_{10min}) \quad (2.2)$$

where m_{15min} is the water absorption at 15 minutes (g); m_{5min} is the water absorption at 5 minutes (g); A is the contact area between the pipe and the earthen surface (m); t_{10min} is equivalent to 10 minutes (s).

The graphs (Fig. 2.29 on the left) show the absorption (g) as a function of the time (min), recorded on two samples for each typology. The greater the average slope of the curve, the greater the rate of absorption C_{10min}. The results confirmed the findings of the wettability test.

Erosion Tests

The erosion test was carried out according to New Zealand Standard NZS 4298, with some adaptations for the criteria of evaluation. Both the Geelong method and the pressure spray method were used.

The Geelong test (Fig. 2.31) involved the dripping of 100 mL of water onto the specimen for a duration going from 20 to 60 minutes and the measurement of the final erosion depth (D, mm). The drops fell from an established height of 400 mm and the specimen, placed with an inclination of 50%, was characterized by an area of 75 × 150 mm and a thickness of 30 mm. The surface treatment was applied on the entire surface of the specimen.

The pressure spray test (Fig. 2.32) simulating the driving rain was done on samples of 250 × 250 mm dimensions applied to a wall panel. It consisted of spraying the exposed section of the sample for a period of 1 hour or until the sample was eroded through, with interruptions at 15-minute intervals to record the erosion depth (D, mm). The water jet was projected from a distance of 470 mm with a pressure of about 50 kPa.

The results (Fig. 2.33) demonstrated the best resistance to water erosion of the plaster treated with silane—siloxane product. Indeed, it was the only one left flawless. Beeswax and silicon nanoparticles treatments were eroded by water, and the erosion also reached the body of the plaster behind it in the first 15 minutes of the test. The titania and silica nanotreatment had the worst resistance to erosion.

2.E DATA SHEET: DENSE PURE FOAMS VERSUS NANOFOAMS

This data sheet aims at demonstrating that the addition of nanoparticles on polyurethane foams to be sprayed as external thermal insulation composite system, ETICS (Fig. 2.34), could enhance both their thermal and mechanical performance. However, some durability issues are still to be resolved.

In the literature, there are many studies on the effect of nanoparticles additions on polyurethane foams. They commonly are focused either on thermal or on mechanical or even on chemical-morphological aspect.

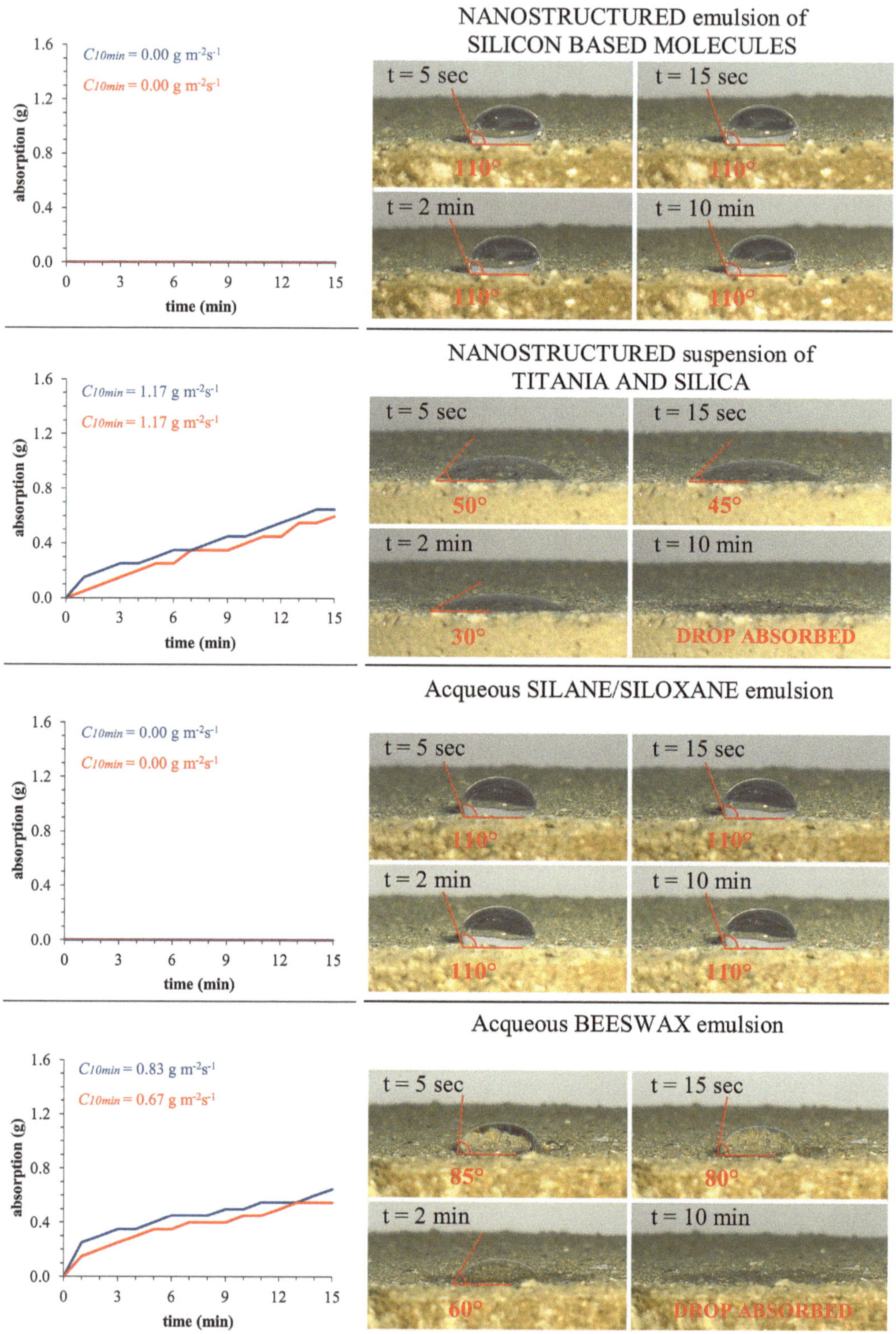

FIG. 2.29 Water absorption and wettability tests results. (From Stazi, Nacci, et al. (2016).)

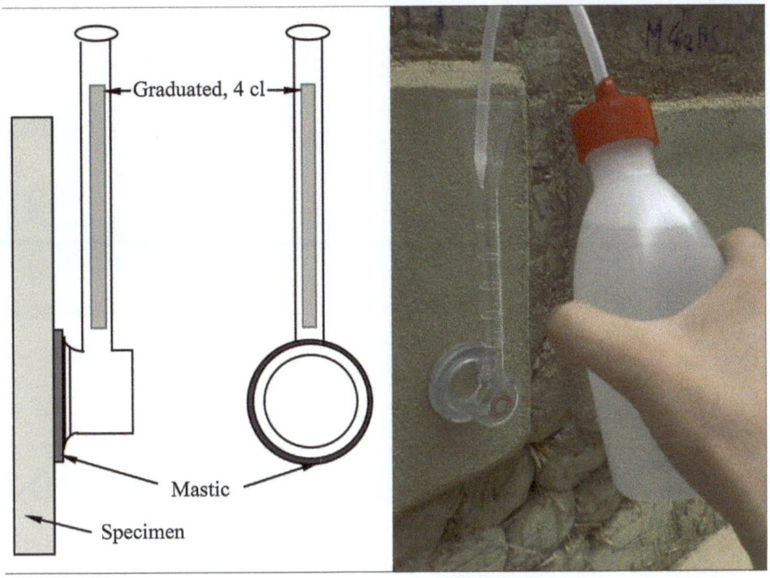

FIG. 2.30 Water absorption test method. (From Stazi, Nacci, et al. (2016).)

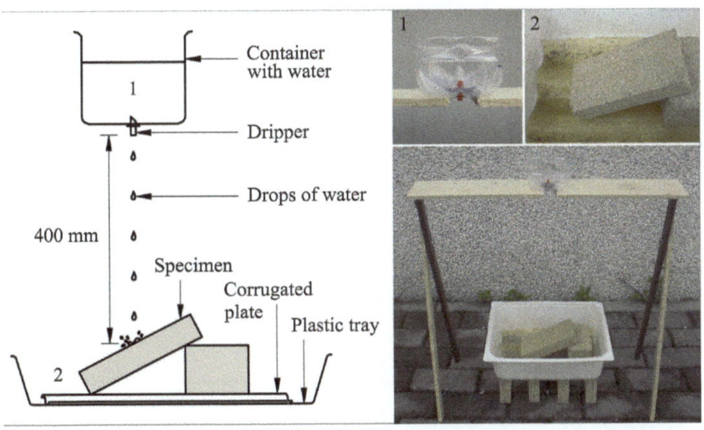

FIG. 2.31 Geelong test method. (From Stazi, Nacci, et al. (2016).)

Rarer are the studies on integrated evaluations of all aspects.

This data sheet reports the results of an experimental campaign aimed at simultaneously comparing the effect of the addiction in polyurethane foams of montmorillonite nanoparticles in different concentrations (2 wt.% and 4 wt.%) and carbon nanofibers in low concentrations (1wt.%). For more details on the methods and adjunctive results, the readers could refer to (Stazi, Tittarelli, et al., 2018) and (Stazi et al., 2019).

In total, nine different panels (50 cm × 50 cm × 7 cm) and three cavity wall samples were manufactured for the thermo-hygrometric and mechanic tests. The panels were then used to cut away the samples with the dimensions requested by the relative standards.

The foam material consisted of two-part polyurethane foam precursors, in the form of liquid, with a nominal density of 30 kg/m^3. In addition, higher densities were tested and some sample was manufactured with an apparent density of 50 kg/m^3. The following nanoparticles were selected:

- Platelet montmorillonite nanoparticles, with thickness of 1 nm, surface of 6–10 μm, and density of 2.5 g/cm^3;

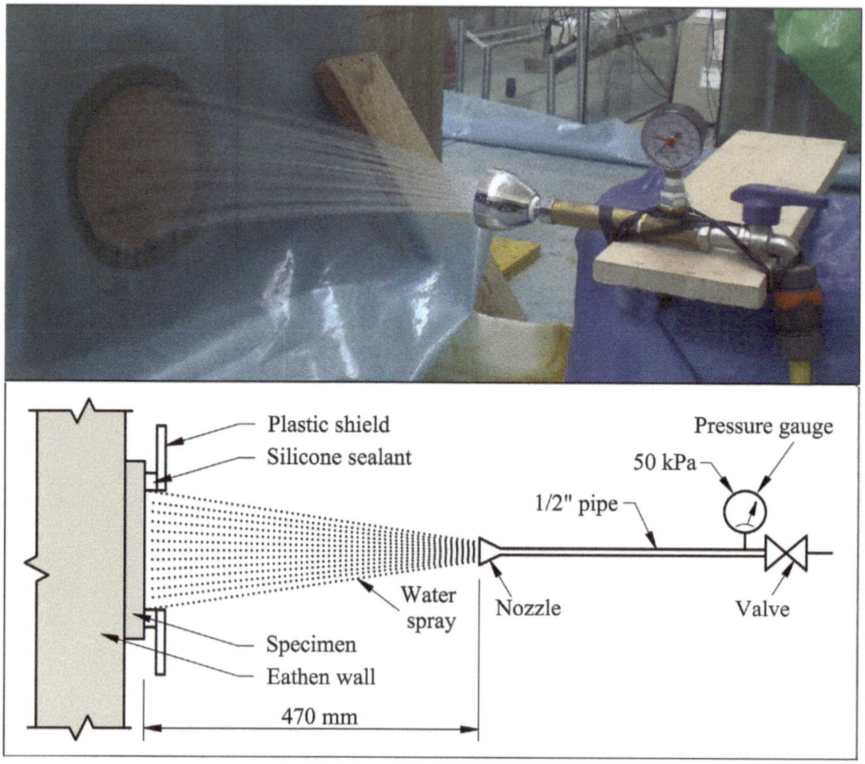

FIG. 2.32 Pressure spray test method. (From Stazi, Nacci, et al. (2016).)

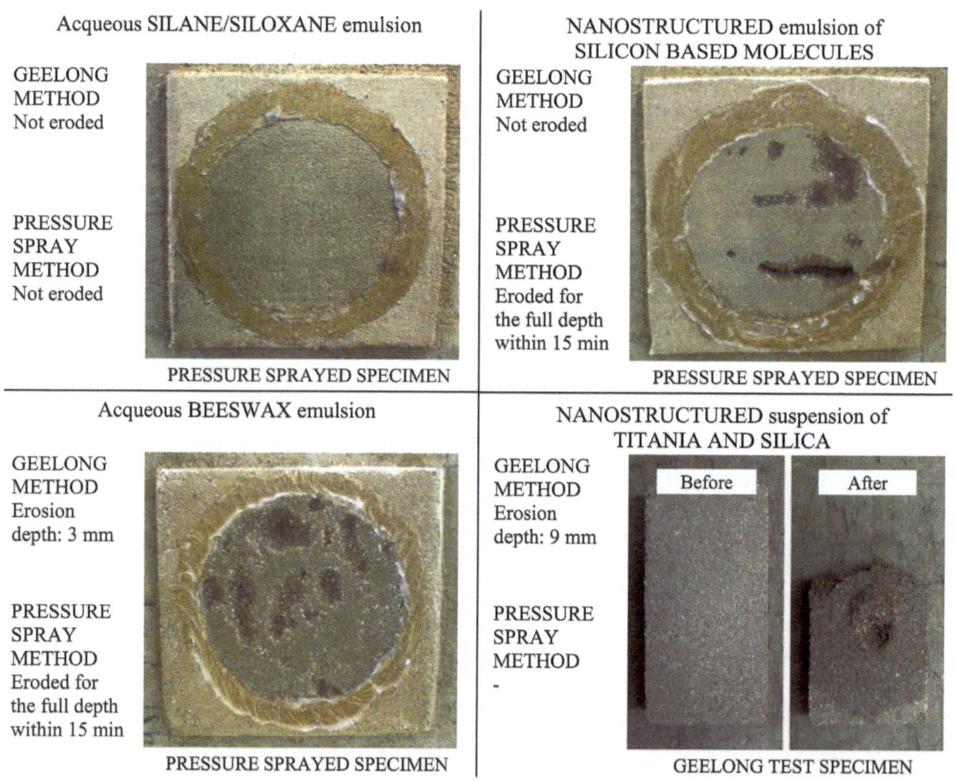

FIG. 2.33 Results of the erosion tests. (From Stazi, Nacci, et al. (2016).)

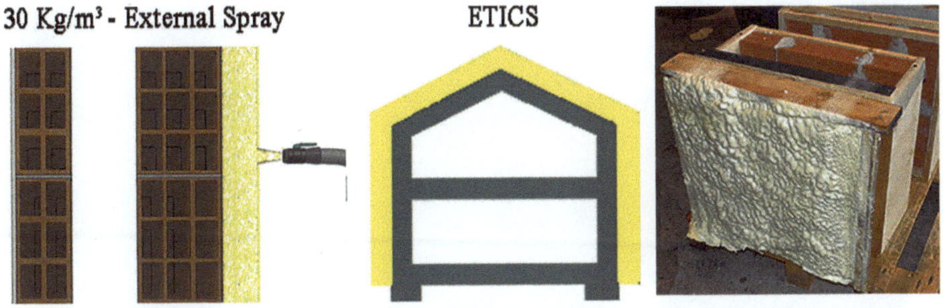

FIG. 2.34 Sprayed polyurethane insulation for external insulation: application schemes and image of assembled components for laboratory tests. (From Stazi et al. (2019).)

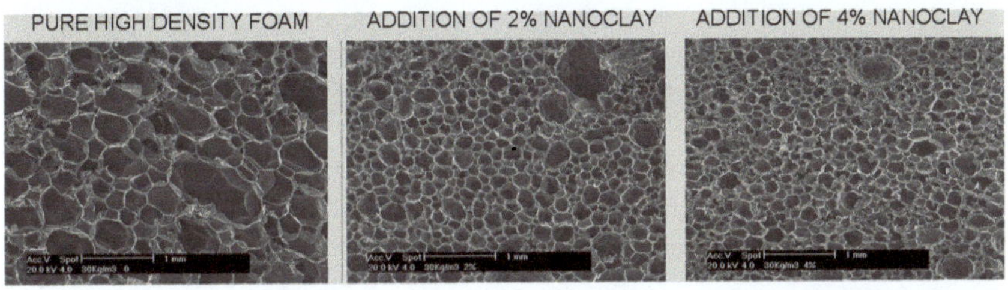

FIG. 2.35 SEM images of pure foams and foams with increasing amounts on nanoclay. (From Stazi et al. (2019).)

- Vapor-grown carbon nanofibers (VGCNFs) with an average diameter in the range of 30–50 nm and a length of 10–30 μm.

In both cases, the nanofibers were homogeneously dispersed into polyol of the PUR foam by an ultrasonic cavitation method.

Subsequently, the nanophased foam was obtained by adding diisocyanate to the mixture and pouring it in a mold. The higher density foam was obtained by the same precursors by adopting a closing cap on the top of the mold, restraining the foam rise. In this case, the surface veil was not removed to reproduce the real conditions of this application in buildings.

Morphology and Density

The overall morphology of neat and nanophased foams was studied using SEM.

The results show that the addition of nanoclay (Fig. 2.35) caused a nucleation phenomenon. Indeed, the nanoparticles behaved as nucleation centers, for the foams determining multiple pores with smaller dimensions. This led to more numerous and smaller cell structures as the amount of nanoparticles increased.

The structure of the neat foams of the two densities (30 and 50 kg/m^3) is characterized by both closed and open cells with variable dimensions, irregular polyhedral shapes, and thin walls (Fig. 2.36). In the case of higher density foams, the structure has smaller cells with thicker walls.

The addition of carbon nanofibers caused in both cases a reduction in cell size, an increase of cell-wall thicknesses, and a more uniform voids distribution. Moreover, the number of cells with broken walls was strongly reduced, especially for nanophased foams with higher densities.

Apparent density test was carried out on prismatic specimens according to UNI EN 1602, before and after the nanoparticles addition (see the test method in Section 4.2.4). The nanoparticles addition increased the densities due to the lower cells size (Table 2.3).

Hydrothermal Performance

The thermal performance was studied through laboratory tests on the panels and wall samples with the heat flow meters method (according to UNI EN 12667) keeping the samples in stationary conditions (see Section 4.2.1). During the test, the specimen was placed between a hot plate (30°C) and a cold plate (10°C) to keep a constant temperature gradient through the specimen. The surface temperature and the heat flow at the center of the samples were continuously measured for 24 hours.

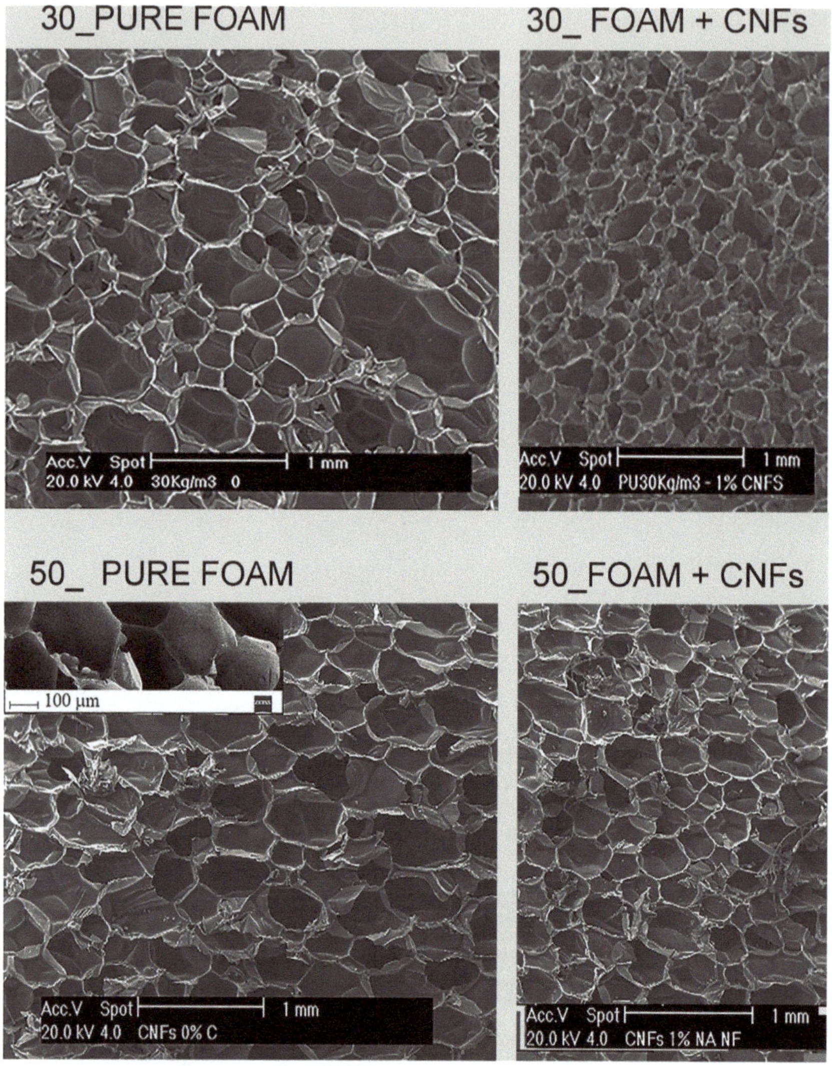

FIG. 2.36 SEM images of foams of different density, both pure and with carbon nanofiber.

TABLE 2.3
Apparent Density Before and After the Nanoparticles Addition.

Sample[a]	ρ (kg/m^3)
30_0%	32.0
30_2%N	32.1
30_4%N	40.2
30_1%CNFS	35.4
50_0%	49.9
50_1%CNFS	50.2

[a] averaged values regarding the apparent density of all samples.

The results (Table 2.4) demonstrate that the regularity increase of the foam structure due to nanoparticle addition determined a reduction of the crossing heat flux and conductivity.

Water vapor permeability was measured according to UNI EN ISO 12572 (see Section 4.2.5). This characteristic depends on open porosity, connected pores size (permeability should be proportional to the product of porosity and square diameter of the main mode pore), connectivity, and tortuosity of the microstructure. In particular, the greater the threshold pore diameter, the higher the permeability of the material.

TABLE 2.4
Thermo-Hygrometric Parameters With and Without the Nanoparticles Addition.

Sample	λ (W/(mK))	U (W/(m^2K)	μ
30_NEAT	0.026	0.35	43.2
30_2%N	0.025	0.34	42.4
30_4%N	0.024	0.33	56.5
30_1%CNFS	0.023	0.32	66.4

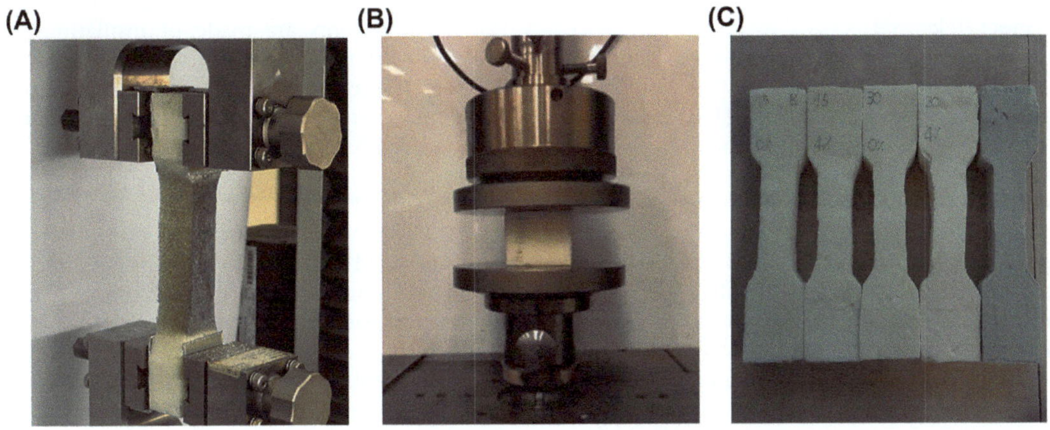

FIG. 2.37 Tensile **(A)** and compressive **(B)** tests, dog bone specimens **(C)**.

The addition of nanoparticles generally increased the hygroscopic resistance factor due to the smaller porous microstructure.

Mechanical Performance

Three samples for each type of foams were considered to determine the average value of mechanical strength and modulus (Fig. 2.37).

Compression tests on cubic specimens were done according to EN 826. The addiction of nanoparticles enhanced the performance with increasing strength values as the amount of the nanoparticles increased (Table 2.5). Sample with low nanoclay dosage made an exception, slightly reducing the σ_{10} value. In 30 kg/m^3 PUR foam, the addition of 1 wt% CNFs increased its compressive strength and Young's modulus by 68% and 140%, respectively; in very high-density foams (50 kg/m^3), the same addition increased its compressive strength and Young's modulus by about 69% and 100%, respectively.

Tensile tests were done according to the ASTMD 638–89 using dog bone–shaped specimens. The tensile strength and modulus were found higher for nanophased foams. For lower density nanoclayed foams, the strength values increased at increasing concentrations with the highest performance for 4% wt. However, the most performing solution was that with carbon nanofibers although in very low quantities

TABLE 2.5
Mechanical Tests Results.

Sample	Ec [kPa]	σc_{10} [kPa]	Et [kPa]	σt_u [kPa]
30_NEAT	932	54	693	48
30_2%N	605	49	1642	126
30_4%N	711	61	1805	65
30_1% CNFs	2245	85	2832	137
50_NEAT	943	91	11873	372
50_1% CNFs	1893	144	15894	413

(note increase in tensile strength and Young's modulus by 277% and 308%, respectively).

In higher density foams, the performance increase in tensile strength and Young's modulus due to the carbon nanofibers was high (by 11% and 34%, respectively) but less noticeable than in softer foams.

Summary and Durability Issue

In general, the carbon nanofibers gave the best improvements.

The nanoparticles addition had greater effects on lower densities foams (30 kg/m^3) with respect to higher density ones (50 kg/m^3).

Our research group also carried out preliminary studies to evaluate whether the improvements are maintained even after aging or not (Pulsoni, degree thesis, 2016). The results revealed that aging had a very positive effect on the pure foam and only a slightly positive effect on the nanoclayed foam, increasing their density due to the achieved higher regularity. Inversely, the foam with CNFs was subjected to a deterioration with voids and broken cells, for the disruption of the reticular regularity and shrinkage, with consequent decrease of both thermal and mechanical performance. So the aging had the effect of inverting the ranking between the foams: the carbon nanofoam was the best before aging while the pure foam achieved the highest performance after aging, even if with high increase in stiffness.

2.F DATA SHEET: VENTILATED MODULE PERFORMANCE

This data sheet aims at demonstrating that new advanced preassembled modules with variable configuration according to the environmental conditions, could be effective solutions for energy saving of both new and existing buildings.

The system presented hereafter was patented by the present author (together with P. Munafò) and involves the use of a preassembled panel with a cost effective substructure realized with the same insulating material.

The anchorage of the panels to the massive support occurs as a normal external insulation, that is, with adhesive and mechanical anchors both placed in correspondence of the spacers. The opening/closing of the ventilation channel takes place through vents positioned in the inferior/superior parts of the walls/roof. The opening vents are made of insulating material equipped with sealing strips and handled by a wireless electronic device with sensors for automatic opening, based on external temperature. The vents are open in summer (ventilated wall) and closed in winter (conservative static wall).

The "special panels" equipped with opening vents (Fig. 2.38) come on-site ready for installation previously provided with the following elements: spacers, vents, and electronic system (powered by electric cables or batteries) for vents opening, expanded metal mesh coupled with insect mesh to protect the openings, and vertical elements of insulating material to avoid thermal bridges (that were removed from the prototype to make the opening visible).

The system was simulated on a building that was also monitored to have an experimentally calibrated model. More details on the case study and simulation method can be found in (Stazi, 2017).

The data (Fig. 2.39 and 2.40) demonstrate that the system is able to reduce the summertime overheating with a lowering of wall internal surface temperatures of a masonry envelope down to 2°C with respect to those recorded with ETICS. The ventilated insulation

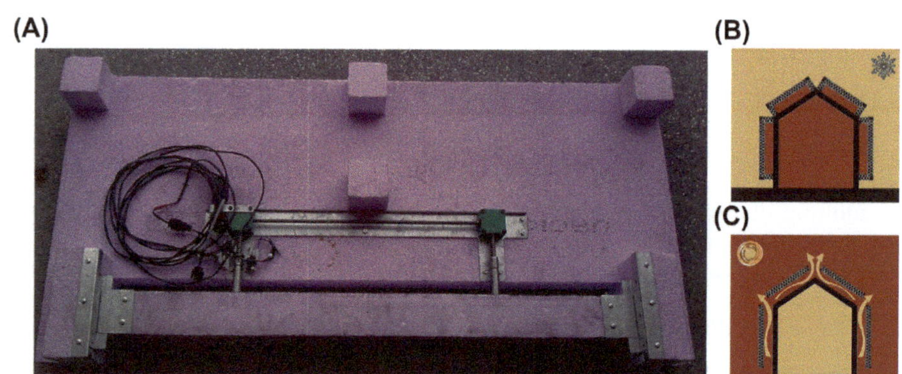

FIG. 2.38 Prototype of the special panel **(A)**, winter **(B)**, and summer **(C)** functioning schemes (Stazi, 2017).

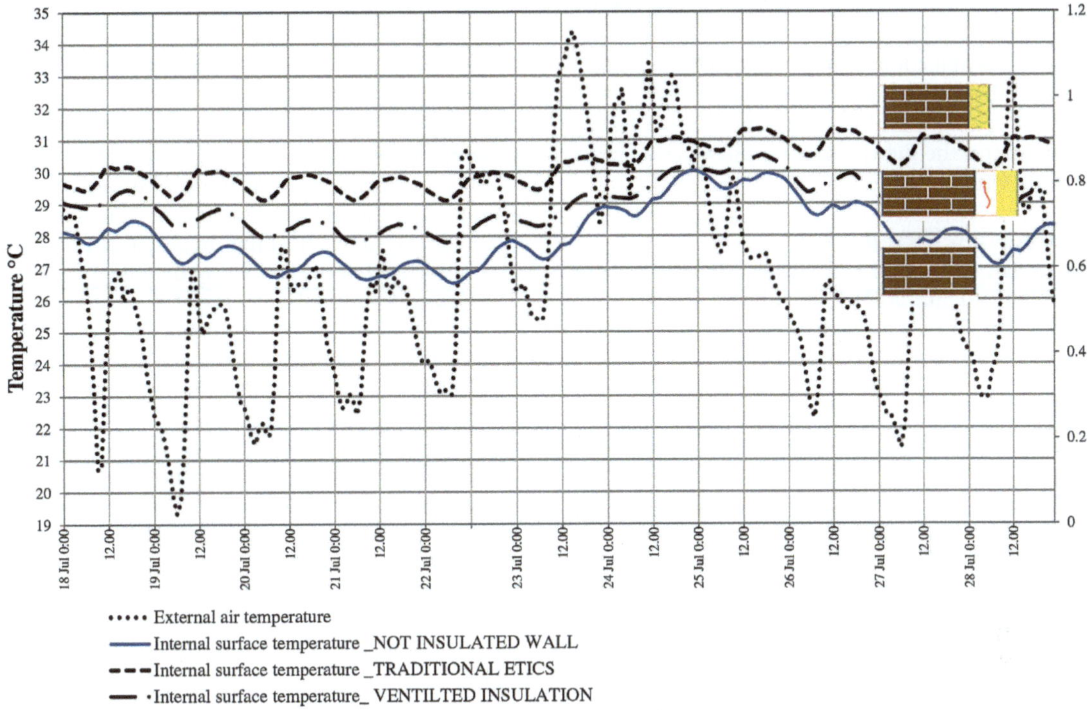

FIG. 2.39 Surface temperatures in not insulated masonry, ETICS, and ventilated module (Stazi, Bonfigli, et al., 2015).

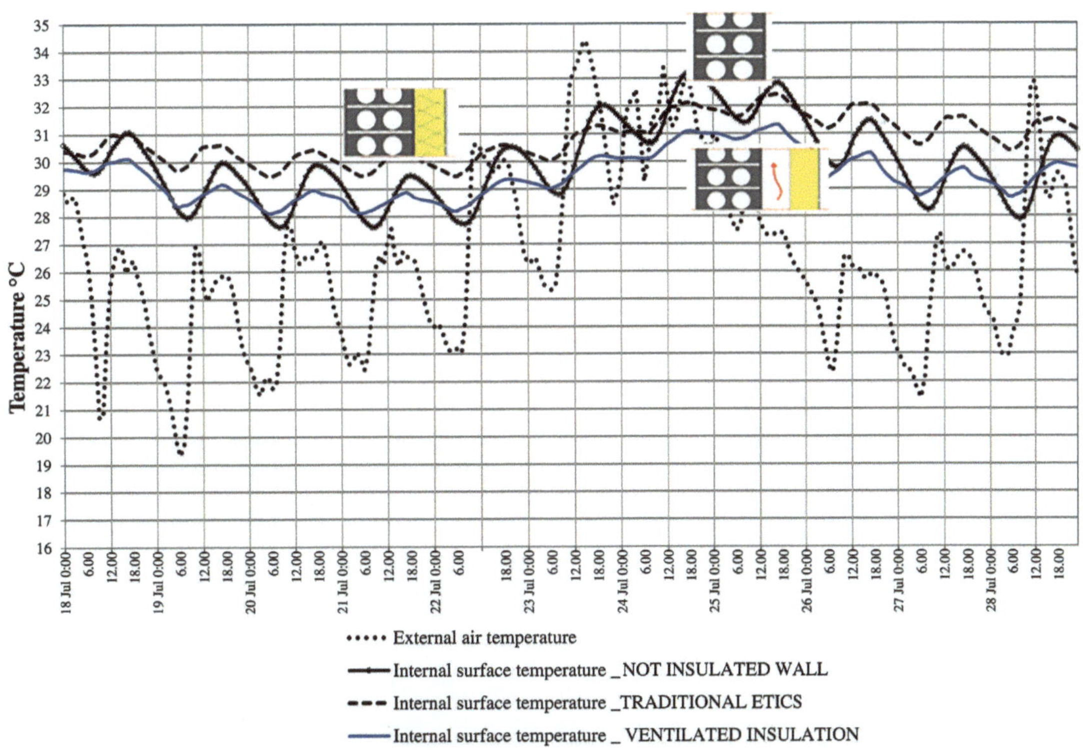

FIG. 2.40 Surface temperatures in not insulated cavity wall, ETICS, and ventilated module.

is also convenient if adopted in perforated brick walls or cement load-bearing panels.

Fig. 2.41 shows a building section with traditional ETIC system (on the right) and the preassembled module (on the left).

2.2.4 Dynamic Design

Recent studies consider the façade as a living skin that can ventilate, dehumidify, and absorb energy self-efficiently. The most surveyed techniques use changes of material state or optical properties to modulate the environmental responsiveness.

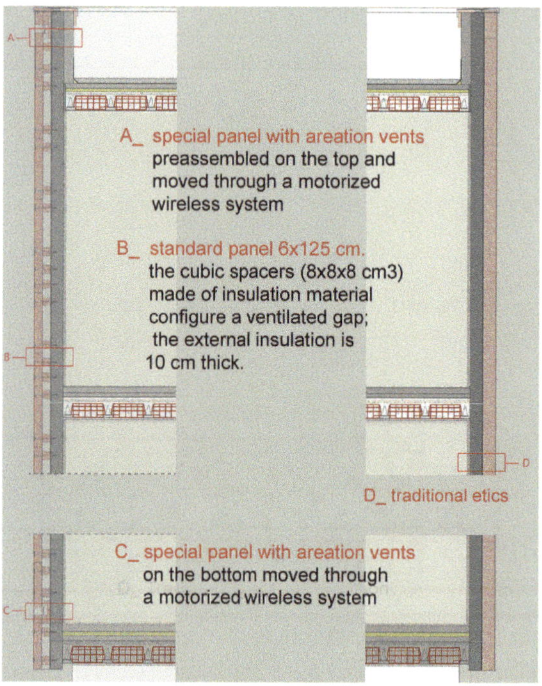

FIG. 2.41 Ventilated module (left) against traditional ETICS (right).

Change of state

The basic principle is using physical phase changes, for example, reaching the melting point or favoring evaporation followed by transpiration or desorption, with the aim of harvesting or dissipating the heat.

The widest studied types are phase change materials (PCM). As the environment overheats and the melting temperature is reached, the phase change material melts and absorbs the excess heat due to its phase change from solid to liquid. This heat is only released when the temperature drops below the specified level and the liquid wax returns to a solid state thereby stabilizing the room temperature. PCM can be used to control the air temperature: contrary to sensible heat storage, latent heat storage occurs at the phase change temperature without a significant raise in temperature.

PCMs are not a new concept, but developments in technology have enabled them to be incorporated within a wide range of existing materials. Moreover, while the traditional use of such system is for indoor temperature control, the application on external paints or roofing cladding systems are more recent. Comparative experimental studies (Fig. 2.42) demonstrated that the integration of PCMs can further improve the performance of cool coatings, thanks to the large heat storage capacity with a nearly isothermal nature of the storage process in which the material undergoes the change of phase (Karlessi et al., 2011). However, further research is necessary to extend the life span of PCM materials and enhance their storage capacity. New highest outcomes of some author also demonstrated the good properties of PCM within wood plastic composites (Chung & Park, 2016).

Building integrated evaporative cooling (BIEC) is an alternative based on phase transition to cool the

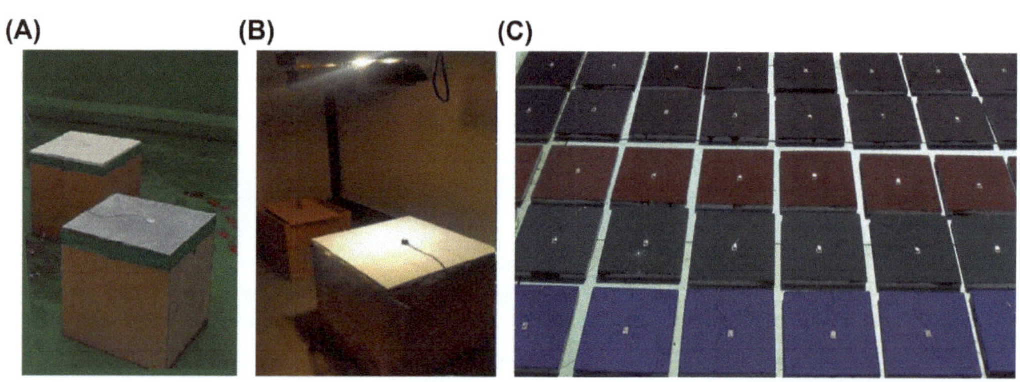

FIG. 2.42 Experimentations on alternative PCM material, by Chung and Park (2016) and Karlessi et al. (2011).

surfaces of a building and to contribute to the external environment quality.

According to this method, if the water sprinkled in a surface or retained within a material pores is evaporated, the material temperature is reduced due to the release of latent heat. The effect is much faster than that of bulk water due to the increased surface area provided by the pores. In turn, the heat flux in the building is reduced due the reduction of material temperatures.

This behavior can be achieved using photocatalyst coatings with sprinkling water or using tailored porous materials with high ability of water vapor adsorption.

In the former case, the water is continuously spread over the surface that should be hydrophilic (i.e., adopting TiO_2) leading to the formation of a very thin layer of water. The building is cooled by a latent heat flux when the water evaporates; however, the continuous water circulation increases the cost of the applications and has greatest difficulties in vertical applications (He & Hoyano, 2008).

In the case of application of highly porous materials, stored water or night adsorbed moisture inside the pores are evaporated during the hot day, and the porous surface temperature is reduced due to the release of the latent heat. Such porous materials have to be superhydrophilic and should have great capacity of water vapor adsorption.

To date, studies have focused mainly on natural materials, such as volcanic ash and siliceous shale or sepiolite (González, Molina-Sabio, & Rodríguez-Reinoso, 2001) that exhibits the highest absorbent capacity of any mineral clay. Moreover, by supporting the TiO_2 photocatalyst on porous hydrophilic materials, multifunctional composites could be produced to enhance the water vapor adsorption, surface solar evaporative cooling, and the degradation of pollutants.

Some author also proposes 3D printed ceramic claddings to enhance evaporation.

The experimental studies mainly regarded laboratory evaporation experiments, such as SEM, water adsorption measurements, and thermal analyses (Karamanis et al., 2012; Li et al., 2017). Some other ones regarded laboratory tests by exposure of small tiles within a wind tunnel (Montazeri, Blocken, & Hensen, 2015) or with the base immersed in a water tank (Sudprasert & Sankaewthong, 2018). Other studies (Fig. 2.43) focused on the experimental evaluations on real-scale mock-ups (He & Hoyano, 2011; Naticchia et al., 2010), in some case also verifying the final impact on the urban environment through numeric evaluations (He & Hoyano, 2008).

Change of optical properties

In this case, a dynamic shifting of the optical properties is operated. For example, thermochromic materials can change their optical and thermal properties in a dynamic way to ameliorate their energy and environmental performance. Indeed, they can change reversibly their color as a function of their surface temperature thus reaching high level of heat absorptivity in winter and high reflectivity in summer. At higher temperatures, they are colorless and reflect solar energy, while at lower temperatures, being colored, they absorb solar energy. This would contribute to the energy saving both in winter and in summer (Santamouris et al., 2011). Comparative experimental tests (Fig. 2.44) demonstrated the superior performance of such coatings with respect to common and cool materials.

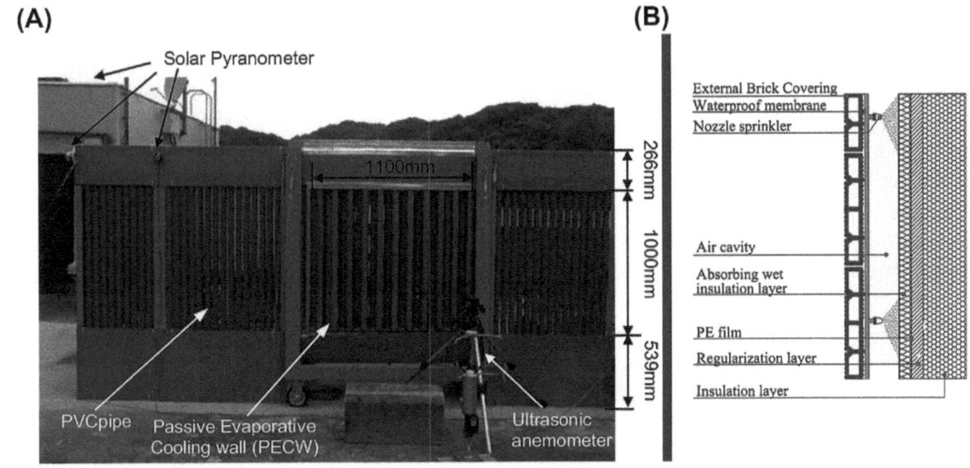

FIG. 2.43 Experiments on passive evaporative cooling walls (He & Hoyano, 2010; Naticchia et al., 2010).

FIG. 2.44 Experimental comparison between common, cool, and thermochromic materials. (From Santamouris et al., 2011).

2.3 INTERNAL SKIN

The present section is about the great importance of enhancing the heat storing ability of the innermost layer of the building envelope that will act as a thermal buffer thus shifting the heat release at later hours.

Thermal energy is stored as either specific or latent heat.

2.3.1 Enhancing the Areal Heat Capacity

The greatest part of the literature focused on the optimal wall configuration agreed on the importance of leaving as high mass as possible on the internal side. However, this measure could turn out to be inefficient if combined with a hyper-insulated wall, especially in warm and temperate climates (Stazi, 2017).

Indeed, such envelopes with thick insulations act as thermal barriers, blocking both the ingoing and outgoing heat flows. Therefore, the high mass positioned inside would not be able to dissipate outward the stored heat in proper time.

To achieve an optimal dynamic behavior of the wall in very high thermal resistance envelopes, the layers have to be selected based on two conditions:

1. The internal mass should be as high as possible (very high internal areal heat capacity, κ_1). This allows the lining to store the heat and release it in a delayed time.
2. The external layers should reduce the crossing fluxes (outgoing in winter and ingoing in summer) without bringing to the extreme the thermal barrier effect.

The latter property implies a precise selection of the decrement factor value, f. This parameter represents the reduction of the periodic temperature oscillations amplitude across the fabric because of heat storage within its inner layers. High attenuating attitude is very important in winter as the heat transfer rate is attenuated for the heat trapped in the internal layers. Part of this heat is also retransmitted inside. Moreover, high attenuating attitude is also positive in summer to dampen the heat coming from outside. However, in the hot season, it could have a negative effect for the heat coming from the inside (due to internal gains). In such period (Fig. 2.45), the greatest amount of heat gains comes from the internal side for the presence of occupants, equipment and, above all, solar gains coming from the windows. A very high damping attitude determines a higher quote of heat trapped inside the wall, which is in part reemitted inside.

Thence, excessively low decrement factor values are to be avoided.

The optimal thermal decrement factor values (synthetically reported hereafter in Table 2.6) are presented and explained in detail in (Stazi, 2017). An f value selected in a range between 0.04 and 0.08 combined with a high internal inertia of the inner lining ($\kappa_1 \geq 40$ kJ/m^{2}K) will configure the best solution.

Data Sheet 2.G experimentally compares alternative internal linings with different combinations of κ_1 and f.

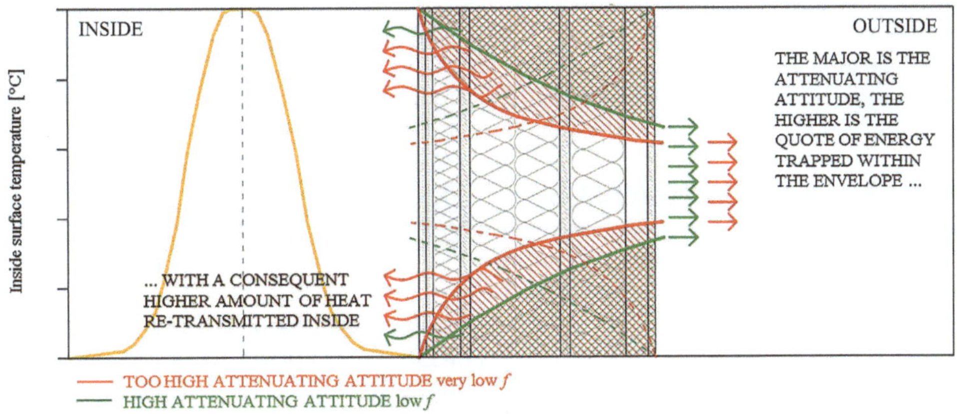

FIG. 2.45 Decrement factor: ideal and excessive attenuating attitude levels. (From Stazi, Tomassoni, & Di Perna, 2017).

TABLE 2.6
Proposal of Combined Limits of Dynamic thermal Parameters, Valid for Energy Efficient Envelopes With U

	Decrement factor f (−)	Internal Areal Heat Capacity κ_1 (kJ/(m²K))	Wall Behavior
1.	$f < 0.04$	$\kappa_1 \geq 50^a$	Thermal barrier
2.	$0.04 \leq f \leq 0.08$	$\kappa_1 \geq 40$	Optimal Solution
3	$f > 0.08$	$\kappa_1 \geq 50$	Crossing fluxes

[a] To be increased up to 60 for $U < 0.2$ W/(m²K).

2.3.2 Enhancing the Latent Heat Capacity

Thermal energy is stored as either sensible or latent heat.

In the former case, the temperature of the medium changes during storage charge or discharge; whereas, in the latter case, the temperature of the medium remains more or less constant as it undergoes a phase transformation. Heat is absorbed and released when the material changes from solid to liquid and vice versa. Thus, it works as latent heat storage unit (read also Section 2.2.4). The thermal capacity of the phase change material (PCM) is not constant but rather concentrated in a temperature range close to the melting point. For this reason, it is essential to choose an appropriate temperature of the phase change so that it is as close as possible to that of comfort to take full advantage of the thermal capacities of these materials.

When the temperature is higher than the melting point, the phase change material is liquid and when the temperature is lowered, it returns to the solid phase. PCMs store energy during their phase transition thus behaving as thermal regulating elements that are able to reduce the daily fluctuations of the ambient temperature. The introduction of phase change materials into building components can significantly increase their thermal inertia, without a substantial increase in weight or volume.

Any PCM composite will comprise two components: a chemical, organic or inorganic, compound that undergoes a phase transition within a desired operating temperature range, and a porous structure that acts as a containment for the heat storage substance. In the building context, pure organic or inorganic PCMs can be encapsulated into the porous structure of traditional construction materials such as gypsum, concrete, or ceramics that are normally used as an internal surface lining.

Different organic (e.g., paraffin, fatty acids) and inorganic (e.g., salt hydrates) chemical compounds have been compared in the literature demonstrating that organic materials are thermally more stable and suitable for encapsulation.

Data sheet 2.H experimentally compares a traditional internal lining (plasterboard) with a clay panel with microencapsulated PCMs.

2.G DATA SHEET: CLAY PANEL VERSUS BRICK PANEL

This data sheet aims at demonstrating that the correct choice of the internal envelope layers (by enhancing the areal heat capacity) is able to reduce the internal surface temperature peaks down to 2°C in summer, thus guaranteeing summer and winter energy saving.

There are many analytical studies on the optimization of dynamic performance of the envelope with respect to indoor comfort and energy saving. Latest studies have demonstrated that, in new super-insulated envelopes, the adoption of a massive internal finishing layer can reduce the discomfort. However, as the following data demonstrate, the optimal solution is not always the one with the highest internal mass.

This data sheet is aimed at experimentally comparing different walls with internal layers (Fig. 2.46) characterized by increasing mass and, inversely, decreasing decrement factor. The optimal combination is experimentally identified. More results are reported in (Stazi, Ulpiani, et al., 2018). The experimental study was carried out in a test room, representative of a full-scale indoor environment at Agugliano, Central Italy (mock-up description in Section 4.4.1). Four internal linings were simultaneously compared.

The details of adopted layers and the thermal parameters for each wall are reported in Table 2.7.

THE FIRST LINING (L) IS LIGHTWEIGHT PLASTERBOARD.

It is a traditional plasterboard finishing 1.25 cm thick; it is also characterized by low thermal conductivity ($\lambda = 0.2$ W/mK), low specific heat capacity C_p and low density, ρ. Thus, the wall with this lining will have a low internal inertia, as the low κ_1 value suggests.

The Second Lining (M) Is Massive, Namely a Clay Panel.

It is a panel of dry clay, 2.8 cm thick; even if it has high thermal conductivity λ and high density ρ it is characterized by a thin section. Thus, it will have a medium inertia (medium k1 value).

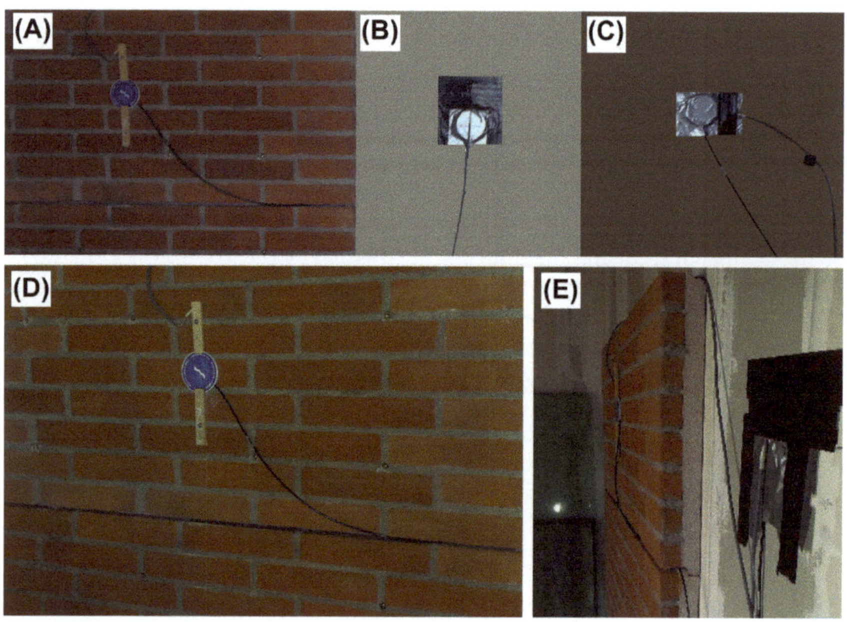

FIG. 2.46 Images of the experimented alternative linings: bricks **(A)**, plasterboard **(B)**, clay panels **(C)**, bricks + insulation **(D, E)**.

TABLE 2.7
Thermal Parameters of the Experimented Walls.

		Low Capacity L	Medium Capacity M	High Capacity H	High Cap + INS H_I
		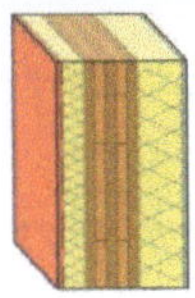	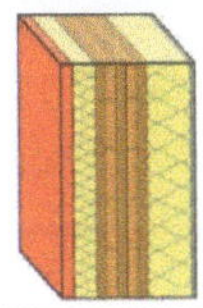	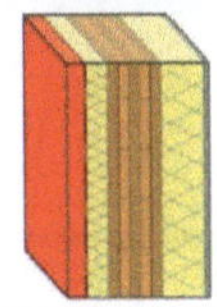	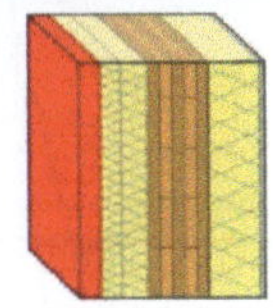
λ	(W/mK)	0.2	0.47	0.47	0.398
C_p	(J/kgK)	837	1000	1000	1000
ρ	(kg/m^3)	760	1300	1300	1491
U	(W/m^{2}K)	0.22	0.22	0.18	0.179
f	(–)	0.076	0.072	0.065	0.047
Δt	(h)	8.39	9.46	10.23	11.02
Y_{12}	(W/m^{2}K)	0.017	0.016	0.014	0.0008
κ_1	(kJ/m^{2}K)	12	33	44	47
T	(cm)	29.5	30.8	32.2	36

The Third Lining (H) Has High Mass, It Is Realized With a Double Clay Panel.

It is realized with two adjacent clay layers, reaching a thickness of 4.2 cm; the other characteristics are the same of lining M, so thanks to its higher dimensions it will reach higher thermal capacity (higher k1 value).

The Fourth Lining (H_I) Is Composed by High Mass (Brick) Plus Insulation.

It is composed by a 4-cm-thick layer of insulation combined with a 4-cm-thick brick layer. The bricklayer has similar features than clay panels but the addition of insulation determines the reduction of U value of the wall and a significant reduction of the decrement factor.

The values of U (W/m^{2}K), f, Δt (h), Y_{12} (W/m^{2}K) and κ_1 (kJ/m^{2}K) are obtained according to EN ISO 13786.

The walls present similar steady state thermal transmittance U, around 0.22 W/m^{2}K; whereas, the addition of a further insulation layer on the internal side of the wall (as occurs for lining H_I) significantly increases the thermal resistance of the wall (therefore lowering the U-value to 0.179 W/m^{2}K). The walls exhibit four increasing levels of internal areal heat capacity, from a minimum of 12 kJ/m^{2}K (lining L) to a maximum of 47 kJ/m^{2}K (lining H_I).

The experimentation provided three tests, representing as many scenarios.

- Test 1: August, 25–30. Medium internal loads (occupancy and equipment) and medium solar heat gains, cooling system on;
- Test 2: October, 9–13. High internal loads and high solar heat gains, heating system on;
- Test 3: December, 23–27. No internal loads or solar gains, heating system on.

As the test room had no windows to avoid uncontrollable heat gains, realistic indoor gains (occupancy, equipment, solar gains for greenhouse effect through a window) were supplied through electric radiators (see Chapter 4.4 for more details).

The external environmental conditions are reported for the scenarios in Fig. 2.47.

Comparison Between Lining L and M

Low and medium thermal capacity envelopes (with lining L and M) have comparable values of decrement factor (around 0.07) and similar time lags but substantially different levels of internal areal heat capacity κ_1, with lining M almost triplicating the value of lining L. They behaved differently in the three tests (Fig. 2.48), but the mass effect could be appreciated especially on test 2, characterized by high indoor gains and low external temperatures (allowing the mass to be cooled at night).

TEST1. During the warmest season, the outdoor temperatures were very high and the internal

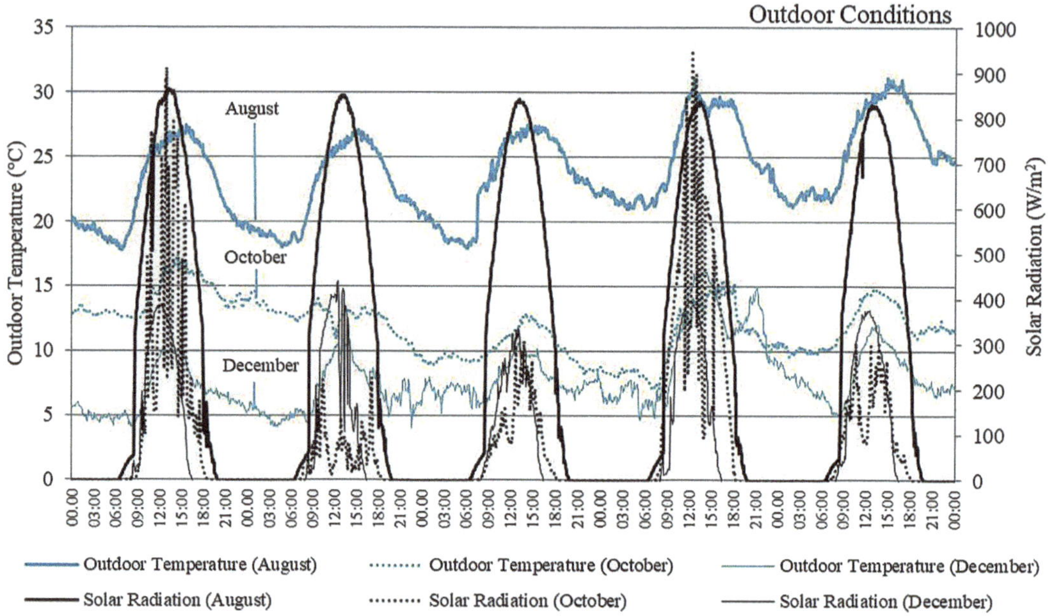

FIG. 2.47 External temperature and solar radiation levels on the three selected periods.

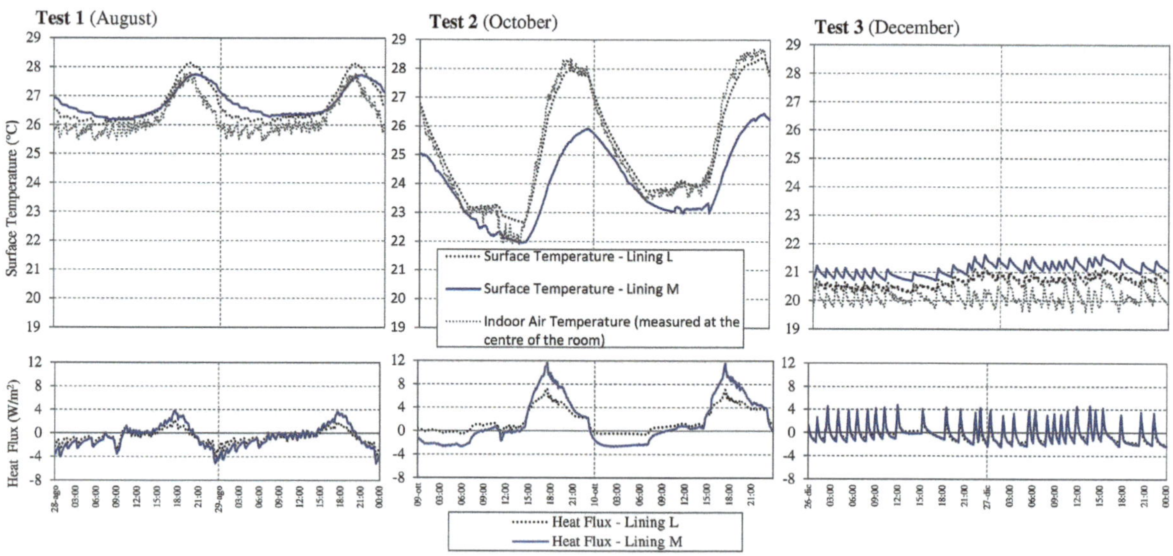

FIG. 2.48 L versus M

conditioning system was switched on (set-point fixed at 26°C). The internal surface temperatures of both linings were elevated throughout the day, varying between 26°C and 28°C. Lining M showed milder thermal fluctuations than lining L. The thermal fluxes have alternate directions at night and during the hottest hours. The highest quotes were recorded as expected by the massive layer, indicating its higher storingand releasing capacity in daytime and nightime, respectively. This demonstrated that medium thermal capacity envelopes could delay and mitigate the peak temperature on the wall surface thus determining a more stable operative temperature.

TEST 2. In October, the external temperatures were decreased but the supplied internal gains were very high, as the solar gains of a sunny summer day were recreated within the test room. In this condition, the mass benefices could be better appreciated, with lining M staying cooler (−2°C in central hours) than lining L. The latter followed the internal temperature trend demonstrating null storing capacity.

The thermal flux for M wall was no more equally distributed between negative and positive fluxes, but moved toward the positive side of the plot for the greater amount of supplied solar gains stored in this internal lining.

TEST 3. In the winter season, the external temperatures drop and the mean surface temperature of Lining M overtook that of Lining L by about 0.5°C. Both surface temperatures and heat fluxes were strongly influenced by the heating system switching on/off.

Comparison Between Linings M, H, and H_I

High thermal capacity wall configurations (with linings H and H_I) have similar values of internal areal heat capacity (over 40 $kJ/m^2 K$) but quite different levels of decrement factor (30% reduction from H to H_I configurations). The wall M is reported as reference.

As in the former comparison, it is evident (Fig. 2.49) that the mass shows its greatest benefices in the intermediate season (Test 2).

TEST 1. Lining M still exhibited the lowest surface temperature, over day and night. Lining H followed the same trend of lining M, keeping constant the reciprocal offset. Contrarily, the gap between lining M and lining H_I is driven to zero in the night, when the negative heat flux through the lining H_I sudden increase (high amount of heat released indoor). This suggests that the interposition of an extra insulation layer behind the internal lining facilitates the heat release in the indoor environment by blocking a deeper penetration of the heat wave thus hindering the expulsion toward outside. The too low f value does not allow to expel toward the outside the heat so that the absorbed quota is totally reemitted inside, causing overheating.

TEST 2. All the observations drawn for test 1 are confirmed under high internal loads conditions, with an even more noticeable difference between M and the other two linings. It is worth noting that the increase of the mass is not followed by the reduction of indoor surface temperatures (as instead occurred moving from L to M in Fig. 2.48). This is because both H and H_I are

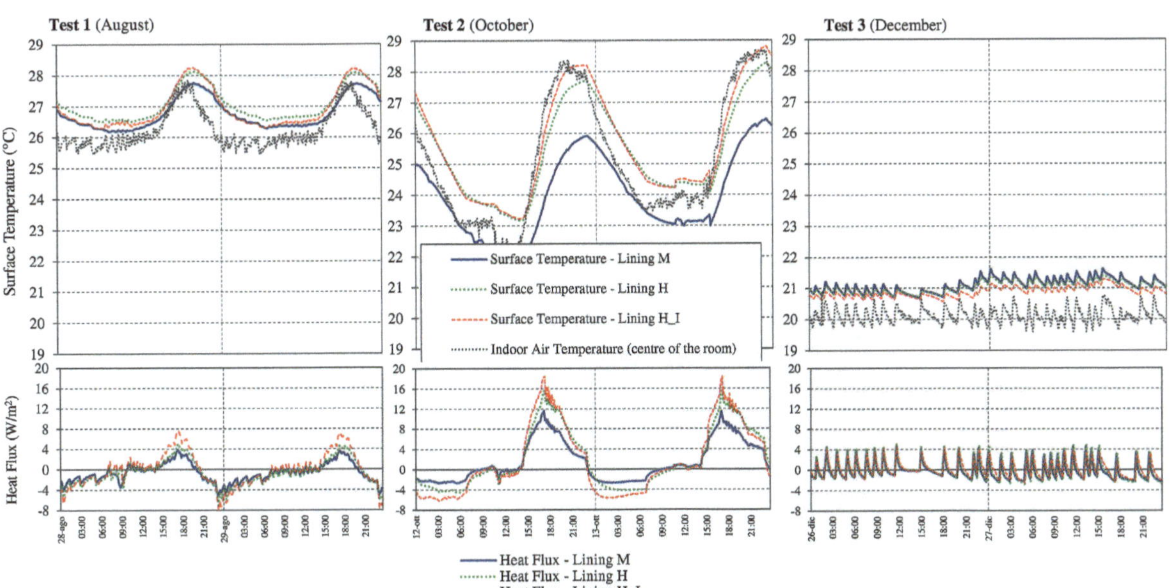

FIG. 2.49 M, H, H_I comparison.

characterized by a considerably reduced value of decrement factor, thus blocking the heat inside and preventing its expulsion outside.

TEST 3. In wintertime, the M solution gave the best outcomes, with slightly higher temperatures.

2.H DATA SHEET: PLASTERBOARD VERSUS PHASE CHANGE MATERIAL

This data sheet aims at demonstrating that phase change (PCM) materials are a low environmental impact technique (enhancing the latent heat capacity) able to guarantee energy saving.

An experimental study was conducted on the comparison between a traditional internal plasterboard and a clay panel with microencapsulated PCMs trapped by a honeycomb structure. Extensive results are reported in (Roccheggiani, degree thesis, 2012).

The experimental study was carried out in a test room, representative of a full-scale indoor environment at Agugliano (see Section 4.4.1).

From the analysis of the state of the art, the paraffin wax was selected as the most suitable for use in constructions for its high thermal stability, especially on the long term. As for the melting temperature, the choice was made by analyzing the average summer temperature of the experimentation site that in summer is around 23.3°C; therefore, it was decided to use a PCM with a melting temperature of 23°C.

The panel (Fig. 2.50) consists of a 4-mm-thick wooden fiber support (1) and a corrugated 8-mm-high honeycombed cardboard that hosts the clay (3) with encapsulated micro PCMs; the panel is finished with a shaving based on clay with sand and supported by a fiberglass mesh (Fig. 2.51).

The total thickness of the panel is 16 mm, with a total weight of 13.5 kg/m^2 of which 3 kg/m^2 are due to the PCM. The thermal parameters declared by the manufacturer are the following: Cp = 1889 J/(KgK); ρ = 1000 kg/m^3. The PCM material consists in aqueous dispersion of paraffin wax encapsulated in highly cross-linked polymethacrylate.

Preliminary Tests

A DSC test was conducted in different samples with about 10 mg of the panel after being shredded and powdered. The heating cycles went from 0°C to 40°C at a heating rate of 2°C/min. The curve of the sample heating is that located in the negative part of the graph while the one placed above represents the cooling curve.

To verify the durability of the material 100 aging cycles were done (Fig. 2.52). Although the heating curves tend to slightly increase the stored energy, no significant changes were detected.

The graph shows melting and freezing onset temperatures of 17.5°C and 22°C, respectively.

FIG. 2.51 Image of the experimented sample.

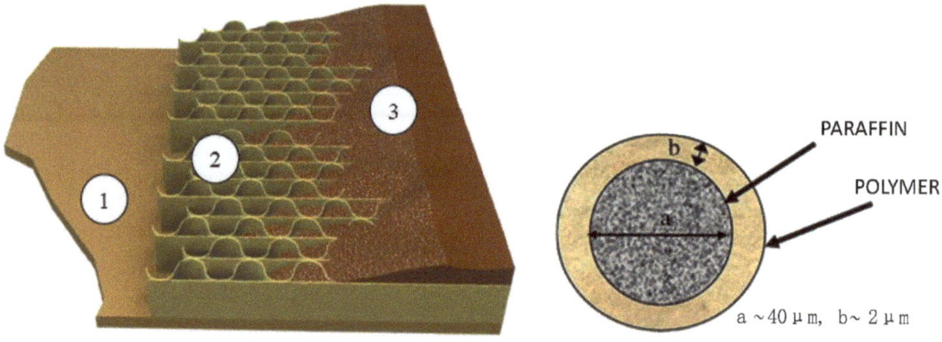

FIG. 2.50 Encapsulated phase change material in a clay panel.

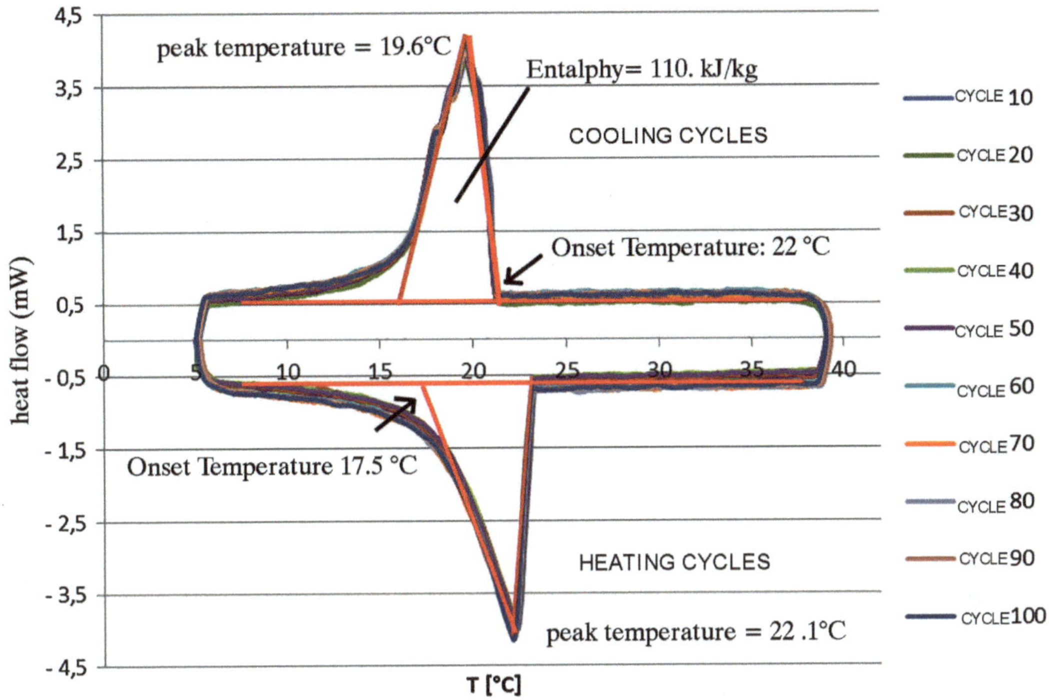

FIG. 2.52 DSC results.

TABLE 2.8
Surface Temperatures, Fluxes, and Thermal Conductivity of the Material in Different Conditions.

Plates Temp.	Surf. Temp. (°C)	Flux (W/m^2)	λ W/(mK)
HOT = 20°C; COLD = 5°C	12.9	27.7	0.103
HOT = 27°C; COLD = 20°C	24.3	3.98	0.038

Laboratory tests on thermal conductivity were done on panels samples (50 × 50 × 1.6 cm^3) with a device based on the heat flow meter method (UNI EN 12664) keeping the sample in stationary conditions. Different temperature values were tested. During the test, the specimens were placed between a hot plate and a cold plate to keep a constant temperature gradient through them.

The surface temperature and the heat flow at the center of the samples were measured for 24 hours. Table 2.8 reports the tested temperatures on hot and cold plate and the relative surface temperature and average fluxes on PCM, with the calculation of thermal conductivity.

The Experimental Study On-Site

In the hot period (July) the indoor temperatures were always higher than the melting point. The comparison with traditional internal plasterboard (Fig. 2.53) shows that PCM gives high benefices, with similar surface temperatures but higher storing ability (higher fluxes towards the wall in the hottest hours).

In winter, with a constant heating set point of 21°C (Fig. 2.54), the use of clay-PCM panels leads to less fluctuating surface temperatures.

Even if the presented on-field tests are only on a preliminary phase, the data suggest that lightweight constructions optimized through the use of internal phase change materials microencapsulated in massive linings should be encouraged (Pajek et al., 2017).

2.4 ENVELOPE CORE

The envelope core is the central cavity within the external wall. It is not present in load-bearing masonries or cross-laminated timber structures. Instead, cavity walls used as infills for concrete frames are conceived with two skins separated by a hollow space. The two

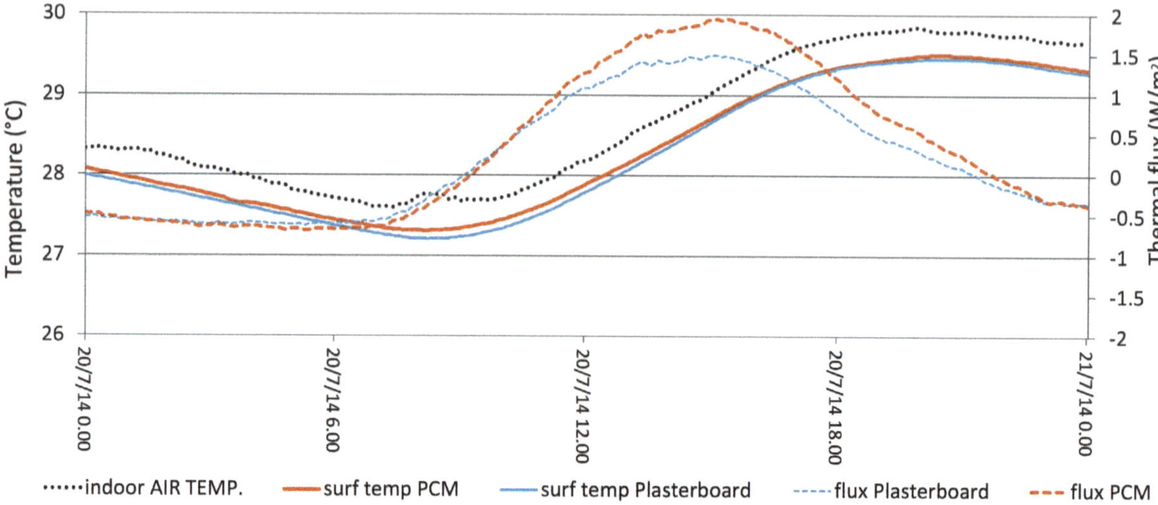

FIG. 2.53 Summer surface temperatures and fluxes of PCM and plasterboard.

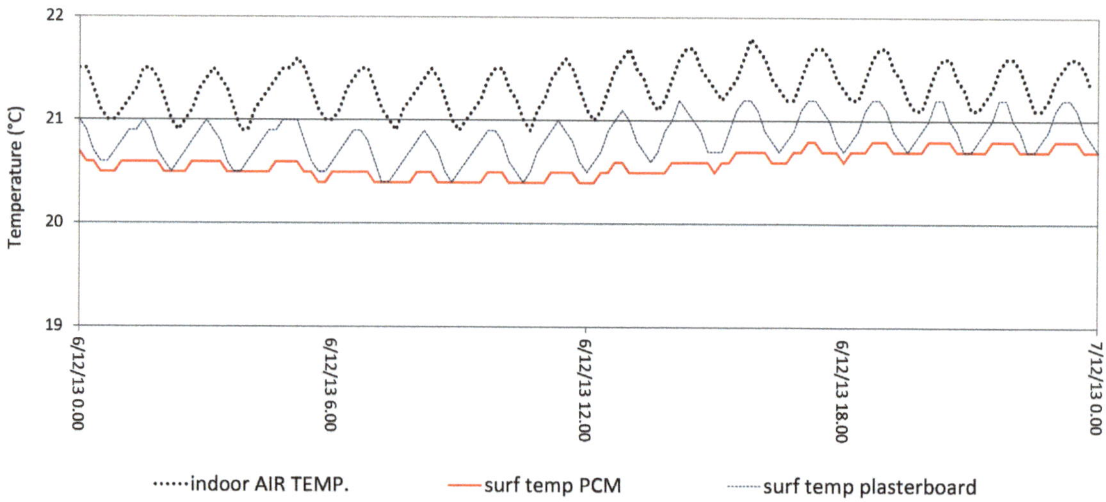

FIG. 2.54 Winter surface temperatures of PCM and plasterboard.

skins are jointed together through stainless steel wall ties at adequate intervals.

The thickness of outer and inner skins may be equal or the former may be thicker than the latter. Cavity wall gap commonly does not exceeds 100 mm and it is uniform through the height. The masonry skins of a cavity wall can be made from bricks or concrete blocks. The cavity of old buildings, realized before insulation standards, is commonly empty, but can be filled with insulation by various methods. The standard method is to drill holes into the outer brickwork at intervals of 1–1.5 m. Insulating beads, granules, loose fiber or foam are then blown into the cavity walls. For retrofit interventions, expandable foams are the most commonly used.

Most spray polyurethane foams are "two-component." Two ingredients—conventionally isocyanate and polyol resin—are mixed on-site using special equipment mounted in a trailer or truck. Heated hoses convey the components to a mixing gun that sprays the chemicals.

A reaction begins as soon as the chemicals are mixed. The liquid mixture foams, expands, and eventually hardens.

For infilled applications, the low density open-cell spray foams are more suitable for cavity wall applications to avoid undesired stress on the external skins and because of the higher permeability to vapor and moisture with respect to high density formulations. Due to its relatively large cell structure, low-density foam remains soft and flexible after curing.

2.I DATA SHEET: SOFT PURE FOAMS VERSUS NANOFOAMS

This section aims at demonstrating that the addition of nanoparticles on soft polyurethane foams to be infilled in cavity walls (Fig. 2.55) could enhance both their thermal and mechanical performance. However, durability problems are still unresolved.

In the literature, there is a lack of studies concerning an integrated evaluation of both mechanical and energy saving aspects. Moreover, very rarely the studies were focused on low-density foams for building sector uses. The main body of the research regards very high-density foams for aerospace or automotive sectors (around 250 kg/m^3). For cavity walls is instead necessary to adopt a soft foam (about 15 kg/m^3) that gives no instability problems for the out of plane movements of the external bricks layers due to internal pushing during the foaming process.

The present data sheet reports the results of an experimental campaign aimed at simultaneously comparing the effect of the addiction of montmorillonite and aerogel microparticles in polyurethane foams in different concentrations. For more details on the methods and adjunctive results, the readers could refer to (Stazi, Tittarelli, et al., 2018).

In total, nine different panels (50 × 50 × 7.2 cm^3) and three cavity wall samples (Fig. 2.56) were manufactured for the thermo-hygrometric and mechanic tests. The panels were then used to cut away the samples with the dimensions requested by the relative standards.

Morphology and Density

The SEM images (Fig. 2.57) revealed that the lowest amount of montmorillonite (2% nanoclay) determines the reduction of the cell size and the breaking of some cell walls. The addition of 4% of nanoclay determines a very denser and more regular structure. The addition of microparticles of silica dioxide (SiO_2) in both concentrations determines a very irregular morphology with voids and localized densifications.

This has an effect on the apparent density (Table 2.9) measured according to UNI EN 1602 (as for Data Sheet 2.E on high density foams). Although the nominal value was 15 kg/m^3 for pure foam, the recorded one was slightly lower, 14.8 kg/m^3. The density increased for both nanoclay and microsilica additions at increasing rates as the amount increased. Very higher density was recorded with 4% of SiO_2 addition.

Hydrothermal Performance

The insulating property of the foams depends on density, dimension of the cells, orientation, and ratio between open and closed cells.

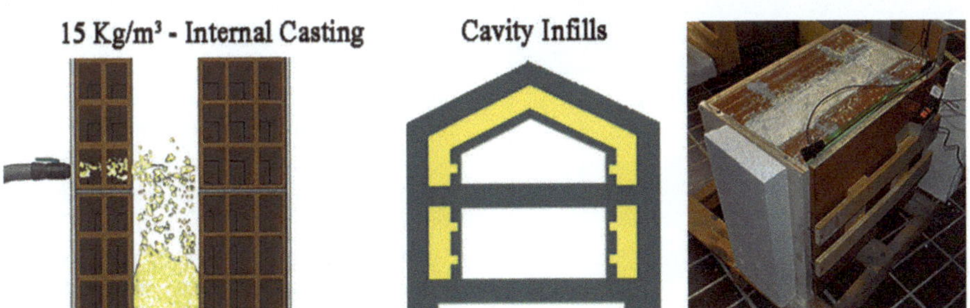

FIG. 2.55 Sprayed polyurethane insulation for cavity wall insulation: application schemes and image of assembled components for laboratory tests.

FIG. 2.56 Cavity walls components for laboratory tests.

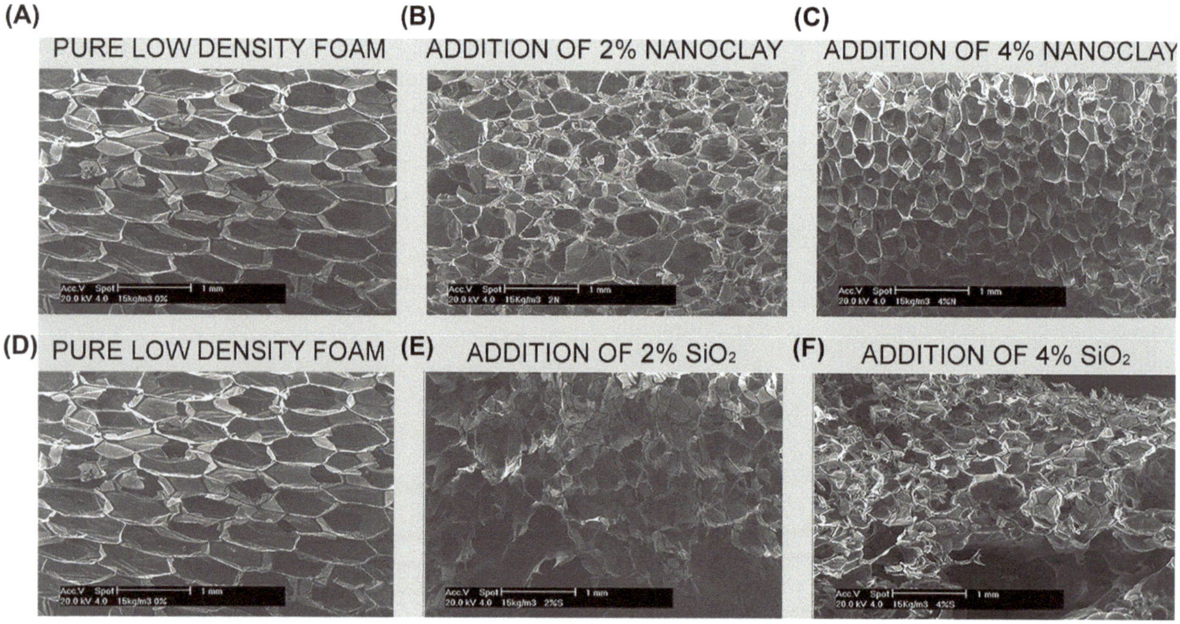

FIG. 2.57 SEM images of pure foams with different densities and increasing amounts of nanoclay and SiO_2.

TABLE 2.9
Apparent Density of the Tested Samples.

Sample*	ρ (kg/m^3)
15_0%	14.8
15_2%N	16.2
15_4%N	19.8
15_2% SiO2	14.2
15_4% SiO2	25.7

The enhancement of the thermal insulating properties due to the particles addition was evident in low-density foams (Table 2.10), with more noticeable differences that those recorded in high density ones for ETICS applications (Data Sheet 2.E).

With nanoclay additions, the structure of the foams has become more regular (especially at high amounts) reducing the crossing heat flux.

Instead, the presence of the aerogel microparticles created an irregular foam structure leading to a conductivity increase even with respect to the neat sample.

TABLE 2.10
Thermal and Hygrometric Performance of the Tested Samples/Components.

Sample	λ (W/(mK))	U (W/(m^{2}K)	μ
15_0% NEAT	0.036	0.3	13.92
15_2% nanoclay	0.034	0.29	18.56
15_4% nanoclay	0.032	0.29	19.51
15_2% SiO2	0.039	0.33	21.56
15_4% SiO2	0.038	0.32	24.85

TABLE 2.11
Mechanical Tests Results.

Sample	Ec [kPa]	σc_{10} [kPa]	Et [kPa]	σt_u [kPa]
15_NEAT	103	8.7	631	32
15_2% Nanoclay	129	8.9	2508	33
15_4% Nanoclay	154	12.5	2749	44
15_2% SiO2	100	7.4	944	24
15_4% SiO2	275	16.5	3048	76

Furthermore, the values of thermal conductance C obtained in experimental phase on real-scale wallets allowed calculating the final assembled wall U value.

The addition of nanoparticles enhanced the hygroscopic resistance factor for the refinement of the microstructure.

Mechanical Performance

The addition of nanoclay enhanced the mechanical performance in both compression and tensile stiffness and strength that considerably increased (Table 2.11). Indeed this addition determined a nucleation phenomenon that led to more numerous and smaller cells as the amount of nanoparticles increased. Instead, the silica dioxide microparticles caused disordered agglomerations of solid material that was able to reach higher strength values only for high amounts. In this case, the filler with microscale rather than nanoscale dimensions negatively influenced the reaction between polyol and diisocyanate so the composites presented irregular composition. In general, the best sample form a mechanical point of view was the one containing the highest percentage of silica dioxide because of highly compact solid material created during the irregular foaming process. However, this could imply difficulties during the realization.

Summary and Durability Issue

In summary considering all the studied aspects, the addition of 4% of nanoclay was found to be the best solution.

Our research group also carried out preliminary studies to evaluate whether the improvements are maintained even after aging or not (Pulsoni, degree thesis, 2016). The results revealed that the aging had similar effects on the two foams, pure and nanoclayed, with increase of the thermal conductivity and compressive/tensile stiffness. The good behavior of nanoclayed foam with respect to pure foam is confirmed, except for a loose in softness with the consequent minor flexibility.

2.5 OPEN QUESTIONS AND FUTURE CHALLENGES

External envelope engineering is gaining attention as advanced studies and new composite technologies are working to transform the building skin in a new membrane, able to interact with the urban environment, buildings occupants, and citizens.

In the international field, many researchers have focused on this challenge.

At the urban level several scientists have demonstrated the need to provide solutions that are able to modulate the albedo by reducing heat peaks, with beneficial effects not only on summer conditioning consumptions but also on materials durability and on human health. These researchers are working on materials and systems able to provide for humidification of urban canyons using water harvesting systems and the reduction of heat through dissipative transformations.

Other advanced solutions to make the envelope more resilient to the environment, propose the envelope as a communication tool entrusting it the interaction with users on both its external and internal side, also with the capacity to learn over time (Velikov & Thun, 2013). The most technological solutions, often only in the design phase, regard kinesthetic interfaces that dynamically change their optic characteristics according to the stimuli and interact with citizens through responsive movements or changes.

Geometry and assemblages could be produced by advanced extrusion or 3D printing technologies. For example, a very interesting solution on evaporative cooling regards 3D printed porous ceramic bricks.

At the building level, the new solutions take to the extreme the energy-saving potential. On this regard, future perspectives envisage (Loonen, Trcka, & Hensen, 2011): making a more smart use of thermal inertia to attenuate peak loads; exploiting knowledge about periodicity of occupancy and load profiles to maximize the benefits of on/off behavior; using weather-based information to extend the period of passive cooling/heating strategies use.

The solutions should rely on simultaneous optimization of several aspects, reuse of materials, form-resisting spatially corrugated structures, integration of nanomaterials and nanodevices, integration of services in modular components, and assembling easiness by adopting dry techniques (De Matteis et al., 2017).

However, the most advanced studies are generally fragmented in various fields of knowledge, chemistry, energy, materials development, structural specialists. There is no synthetic vision for the building sector allowing to combine the most advanced solutions in cost-effective applications. This objective could be facilitated by increased interactions between academia and industry and a targeted funding strategy of national governments.

Another question that needs an answer is the quantification on how such dynamic and new technologies will affect the energy usage and the urban quality.

The efficacy evaluation should move from the experimental quantification on prototypes toward extensive numeric analyses of the real effects on building energy efficiency and urban levels, for example, by identifying the impact of a passive cooling technique on the urban albedo increase and on the consequent decrease of the long-term global temperature.

Moreover, the new solutions with multiple performance (for example, integrated energy-structural approaches) pose the problem to tune evaluation methods able to capture the intervention convenience with respect to traditional decoupled approaches. Firstly, this type of evaluation should consider the real lifetime of the building fabric, that for innovative solutions is still an open question and based on not uniform data. Moreover, energy saving is actually obtained through building performance simulations based on point estimates, while the thermal performance of the envelope strongly depends on parameters linked to the uneasily predictable user behavior on building devices. Such uncertainty makes it difficult to accurately evaluate the energy-saving strategies.

There is a need of new probabilistic assessment methodologies (based, for example, on LCIA, LCC, and CO analysis) that are able to take into account the uncertainties, to include different aspects (thermal, structural, etc.), and to include the effective data regarding the service life.

This would surely increase the evaluation complexities and computing times: problem that could be partially resolved by choosing an appropriate model resolution.

Moreover, a still open problem regards the durability. The effect of aging on new coatings should be further investigated, with regard to the energy saving potential during the whole lifetime of a construction. For example, the addition of nanoparticle to existing materials could change the molecular structure of the binder thus determining long-term reductions of the material performance.

Finally, a very efficient energy-saving measure could create long-term inefficiencies, for example, unaesthetic appearance and thermal performance reduction, which need to be addressed. For example, thick thermal insulations of recent ETICS are more prone to durability problems for biological growth (D'Orazio et al., 2014), as impeding the exit of the heat flux from the inside could cause a temperature drop in the exterior finishing and consequent condensation phenomena.

REFERENCES

Akbari, H., Damon Matthews, H., & Seto, D. (2012). The long-term effect of increasing the albedo of urban areas. *Environmental Research Letters*. https://doi.org/10.1088/1748-9326/7/2/024004.

Alonso, C., Martín-Consuegra, F., Oteiza, I., Asensio, E., Pérez, G., Martínez, I., & Frutos, B. (2017). Effect of façade surface finish on building energy rehabilitation. *Solar Energy, 146,* 470–483. https://doi.org/10.1016/j.solener.2017.03.009.

Andaloro, A., Mazzucchelli, E. S., Lucchini, A., & Pedeferri, M. P. (2017). Photocatalytic self-cleaning coatings for building facade maintenance. Performance analysis through a case-study application. *Journal of Facade Design and Engineering*. https://doi.org/10.3233/FDE-160054.

Bedon, C. (2016). Review on the use of FRP composites for façades and building skins. *American Journal of Engineering and Applied Sciences, 9*(3). https://doi.org/10.3844/ajeassp.2016.713.723.

Borri, A., Corradi, M., Sisti, R., Buratti, C., Belloni, E., & Moretti, E. (2016). Masonry wall panels retrofitted with thermal-insulating GFRP-reinforced jacketing. *Materials and Structures/Materiaux et Constructions*. https://doi.org/10.1617/s11527-015-0766-4.

Bournas, D. A. (2018). Concurrent seismic and energy retrofitting of RC and masonry building envelopes using inorganic textile-based composites combined with insulation materials: A new concept. *Composites Part B: Engineering*. https://doi.org/10.1016/j.compositesb.2018.04.002.

Buratti, C., Belloni, E., Lunghi, L., Borri, A., Castori, G., & Corradi, M. (2016). 'Mechanical characterization and thermal conductivity measurements using of a new "small hot-box" apparatus: Innovative insulating reinforced coatings analysis'. *Journal of Building Engineering*. https://doi.org/10.1016/j.jobe.2016.05.005.

Chen, L., Rende, D., Schadler, L. S., & Ozisik, R. (2013). Polymer nanocomposite foams. *Journal of Materials Chemistry A*. https://doi.org/10.1039/c2ta00086e.

Chen, J., & Poon, C. (2009). Photocatalytic construction and building materials: From fundamentals to applications. *Building and Environment*. https://doi.org/10.1016/j.buildenv.2009.01.002.

Chung, M. H., & Park, J. C. (2016). Development of PCM cool roof system to control urban heat island considering temperate climatic conditions. *Energy and Buildings*. https://doi.org/10.1016/j.enbuild.2015.12.056.

Dalla Mora, T., Righi, A., Peron, F., & Romagnoni, P. (2015). Functional, energy and seismic retrofitting in existing building: An innovative system based on xlam technology. *Energy Procedia*. https://doi.org/10.1016/j.egypro.2015.11.851.

De Castro San Román J. Experiments on epoxy, polyurethane and ADP adhesives. Appendix A-Technical Report no. CCLab2000.1b/1, CCLAB, Ecole Polytechnique Federale de Losanne.

De Matteis, V., Cannavale, A., Coppola, A., & Fiorito, F. (2017). 'Nanomaterials and smart nanodevices for modular dry constructions: The project "easy house". *Procedia Engineering*. https://doi.org/10.1016/j.proeng.2017.04.230.

Dye, S., Kopelove, A., & Mills, G. L. (2010). Integrated and load responsive multilayer insulation. In *AIP Conference Proceedings*. https://doi.org/10.1063/1.3422464.

D'Orazio, M., Cursio, G., Graziani, L., Aquilanti, L., Osimani, A., Clementi, F., Yepremian, C., Lariccia, V., & Amoroso, S. (2014). Effects of water absorption and surface roughness on the bioreceptivity of ETICS compared to clay bricks. *Building and Environment*. https://doi.org/10.1016/j.buildenv.2014.03.018.

Elrayies, G. M. (2018). Microalgae: Prospects for greener future buildings. *Renewable and Sustainable Energy Reviews*. https://doi.org/10.1016/j.rser.2017.08.032.

Escudero, C., Martin, K., Erkoreka, A., Flores, I., & Sala, J. M. (2013). Experimental thermal characterization of radiant barriers for building insulation. *Energy and Buildings*. https://doi.org/10.1016/j.enbuild.2012.12.043.

Fantucci, S., & Serra, V. (2017). Low-E paints enhanced building components: Performance, limits and research perspectives. *Energy Procedia*. https://doi.org/10.1016/j.egypro.2017.08.154.

Feroldi, F., Marini, A., Badiani, B., Plizzari, G., Giuriani, E., Riva, P., & Belleri, A. (2013). Energy efficiency upgrading, architectural restyling and structural retrofit of modern buildings by means of "engineered" double skin façade. In *Structures and Architecture*. https://doi.org/10.1201/b15267-253.

Gallo, P., & Romano, R. (2017). Adaptive facades, developed with innovative nanomaterials, for a sustainable architecture in the Mediterranean area. *Procedia Engineering*. https://doi.org/10.1016/j.proeng.2017.04.289.

Giancola, E., Sanjuan, C., Blanco, E., & Heras, M. R. (2012). Experimental assessment and modelling of the performance of an open joint ventilated façade during actual operating conditions in Mediterranean climate. *Energy and Buildings*. https://doi.org/10.1016/j.enbuild.2012.07.035.

González, J. C., Molina-Sabio, M., & Rodríguez-Reinoso, F. (2001). Sepiolite-based adsorbents as humidity controller. *Applied Clay Science*. https://doi.org/10.1016/S0169-1317(01)00062-X.

He, J., & Hoyano, A. (2008). A numerical simulation method for analyzing the thermal improvement effect of super-hydrophilic photocatalyst-coated building surfaces with water film on the urban/built environment. *Energy and Buildings*. https://doi.org/10.1016/j.enbuild.2007.08.003.

He, J., & Hoyano, A. (2010). Experimental study of cooling effects of a passive evaporative cooling wall constructed of porous ceramics with high water soaking-up ability. *Building and Environment*. https://doi.org/10.1016/j.buildenv.2009.07.002.

He, J., & Hoyano, A. (2011). Experimental study of practical applications of a passive evaporative cooling wall with high water soaking-up ability. *Building and Environment*. https://doi.org/10.1016/j.buildenv.2010.07.004.

Jelle, B. P., Kalnæs, S. E., & Gao, T. (2015). Low-emissivity materials for building applications: A state-of-the-art review and future research perspectives. *Energy and Buildings*, 329–356. https://doi.org/10.1016/j.enbuild.2015.03.024.

Karamanis, D., Vardoulakis, E., Kyritsi, E., & Ökte, N. (2012). Surface solar cooling through water vapor desorption from photo-responsive sepiolite nanocomposites. In *Energy Conversion and Management* (pp. 118–122). https://doi.org/10.1016/j.enconman.2012.01.035.

Karlessi, T., Santamouris, M., Synnefa, A., Assimakopoulos, D., Didaskalopoulos, P., & Apostolakis, K. (2011). Development and testing of PCM doped cool colored coatings to mitigate urban heat island and cool buildings. *Building and Environment*. https://doi.org/10.1016/j.buildenv.2010.09.003.

Langmans, J., Desta, T. Z., Alderweireldt, L., & Roels, S. (2015). Experimental analysis of cavity ventilation behind residential rainscreen cladding systems. *Energy Procedia*. https://doi.org/10.1016/j.egypro.2015.11.290.

Li, H., He, Y., Liu, Z., Jiang, B., & Huang, Y. (2017). A flexible thin-film membrane with broadband Ag@TiO_2 nanoparticle for high-efficiency solar evaporation enhancement. *Energy*. https://doi.org/10.1016/j.energy.2017.07.180.

Loonen, R. C. G. M., Trčka, M., Cóstola, D., & Hensen, J. L. M. (2013). Climate adaptive building shells: State-of-the-art and future challenges. *Renewable and Sustainable Energy Reviews*, 483–493. https://doi.org/10.1016/j.rser.2013.04.016.

Loonen, R. C. G. M., Trcka, M., & Hensen, J. L. M. (2011). Exploring the potential of climate adaptive building shells. In *Building simulation 2011: 12th Conference of international building performance simulation Association*. citeulike-article-id:10500167.

Marinosci, C., Semprini, G., & Morini, G. L. (2014). Experimental analysis of the summer thermal performances of a naturally ventilated rainscreen façade building. *Energy and Buildings*. https://doi.org/10.1016/j.enbuild.2013.12.044.

Martin, S., Brown, P. S., & Bhushan, B. (2017). Fabrication techniques for bioinspired, mechanically-durable, superliquiphobic surfaces for water, oil, and surfactant repellency. *Advances in Colloid and Interface Science*. https://doi.org/10.1016/j.cis.2017.01.004.

Montazeri, H., Blocken, B., & Hensen, J. L. M. (2015). Evaporative cooling by water spray systems: CFD simulation, experimental validation and sensitivity analysis. *Building and Environment*. https://doi.org/10.1016/j.buildenv.2014.03.022.

Naticchia, B., D'Orazio, M., Carbonari, A., & Persico, I. (2010). Energy performance evaluation of a novel evaporative cooling technique. *Energy and Buildings*. https://doi.org/10.1016/j.enbuild.2010.05.029.

Negro, P., & Mola, E. (2017). A performance based approach for the seismic assessment and rehabilitation of existing RC buildings. *Bulletin of Earthquake Engineering*. https://doi.org/10.1007/s10518-015-9845-8.

Pajek, L., Hudobivnik, B., Kunič, R., & Košir, M. (2017). Improving thermal response of lightweight timber building envelopes during cooling season in three European locations. *Journal of Cleaner Production*. https://doi.org/10.1016/j.jclepro.2017.04.098.

Principi, P., & Fioretti, R. (2012). Thermal analysis of the application of PCM and low emissivity coating in hollow bricks. *Energy and Buildings*. https://doi.org/10.1016/j.enbuild.2012.04.022.

Ragesh, P., Anand Ganesh, V., Nair, S. V., & Nair, A. S. (2014). A review on "self-cleaning and multifunctional materials". *Journal of Materials Chemistry, 2*(36), 14773–14797. https://doi.org/10.1039/C4TA02542C.

Saha, M. C., Kabir, M. E., & Jeelani, S. (2008). Enhancement in thermal and mechanical properties of polyurethane foam infused with nanoparticles. *Materials Science and Engineering A*. https://doi.org/10.1016/j.msea.2007.06.060.

Santamouris, M., Synnefa, A., & Karlessi, T. (2011). Using advanced cool materials in the urban built environment to mitigate heat islands and improve thermal comfort conditions. *Solar Energy*. https://doi.org/10.1016/j.solener.2010.12.023.

Sassu, M., Stochino, F., & Mistretta, F. (2017). Assessment method for combined structural and energy retrofitting in masonry buildings. *Buildings, 7*(3), 71. https://doi.org/10.3390/buildings7030071.

Shoukry, H., Kotkata, M. F., Abo-EL-Enein, S. A., Morsy, M. S., & Shebl, S. S. (2016). Thermo-physical properties of nanostructured lightweight fiber reinforced cementitious composites. *Construction and Building Materials*. https://doi.org/10.1016/j.conbuildmat.2015.10.188.

Smargiassi, A., Goldberg, M. S., Plante, C., Fournier, M., Baudouin, Y., & Kosatsky, T. (2009). Variation of daily warm season mortality as a function of micro-urban heat islands. *Journal of Epidemiology & Community Health*. https://doi.org/10.1136/jech.2008.078147.

Stazi, F. (2017). Thermal inertia in energy efficient building envelopes. In *Butterworth-Heinemann*. Elsevier. https://doi.org/10.1016/C2016-0-00641-1.

Stazi, F., Bonfigli, C., Tomassoni, E., Di Perna, C., & Munafò, P. (2015b). The effect of high thermal insulation on high thermal mass: Is the dynamic behaviour of traditional envelopes in Mediterranean climates still possible? *Energy and Buildings, 88*. https://doi.org/10.1016/j.enbuild.2014.11.056.

Stazi, F., Tomassoni, F., Vegliò, A., & Di Perna, C. (2011). Experimental evaluation of ventilated walls with an external clay cladding. *Renewable Energy, 36*(12). https://doi.org/10.1016/j.renene.2011.05.016.

Stazi, F., Ulpiani, G., Pergolini, M., Magni, D., & Di Perna, C. (2018b). Experimental comparison between three types of opaque ventilated facades. *The Open Construction & Building Technology Journal, 12*(1), 296–308. https://doi.org/10.2174/1874836801812010296.

Stazi, F., Urlietti, C., Di Perna, C., Chiappini, G., Rossi, M., & Tittarelli, F. (2019). Thermal and mechanical optimization of nano-foams for sprayed insulation. *Construction and Building Materials, 201*, 828–841. https://doi.org/10.1016/j.conbuildmat.2018.12.177. Elsevier Ltd.

Stazi, F., Giampaoli, M., Rossi, M., & Munafò, P. (2015a). Environmental ageing on GFRP pultruded joints: Comparison between different adhesives. *Composite Structures, 133*. https://doi.org/10.1016/j.compstruct.2015.07.067.

Stazi, F., Giampaoli, M., Nisi, L., Rossi, M., & Munafò, P. (2016). Mechanical performance reduction of GFRP specimens with polyester matrix exposed to continuous condensation. *Composites Part B: Engineering, 99*, 330–339. https://doi.org/10.1016/j.compositesb.2016.05.062.

Stazi, F., Nacci, A., Tittarelli, F., Pasqualini, E., & Munafò, P. (2016). An experimental study on earth plasters for earthen building protection: The effects of different admixtures and surface treatments. *Journal of Cultural Heritage, 17*. https://doi.org/10.1016/j.culher.2015.07.009.

Stazi, F., Tittarelli, F., Saltarelli, F., Chiappini, G., Morini, A., Cerri, G., & Lenci, S. (2018a). Carbon nanofibers in polyurethane foams: Experimental evaluation of thermo-hygrometric and mechanical performance. *Polymer Testing, 67*. https://doi.org/10.1016/j.polymertesting.2018.01.028.

Stazi, F., Tomassoni, E., & Di Perna, C. (2017). Super-insulated wooden envelopes in Mediterranean climate: Summer overheating, thermal comfort optimization, environmental impact on an Italian case study. *Energy and Buildings, 138*. https://doi.org/10.1016/j.enbuild.2016.12.042.

Stazi, F., Ulpiani, G., Pergolini, M., & Di Perna, C. (2018c). The role of areal heat capacity and decrement factor in case of hyper insulated buildings: An experimental study. *Energy and Buildings, 176*, 310–324. https://doi.org/10.1016/j.enbuild.2018.07.034. Elsevier B.V.

Sudprasert, S., & Sankaewthong, S. (2018). Utilization of rice husks in a water-permeable material for passive evaporative cooling. *Case Studies in Construction Materials*. https://doi.org/10.1016/j.cscm.2017.12.005.

Sung, D. (2016). A new look at building facades as infrastructure. *Engineering, 2*(1), 63–68. https://doi.org/10.1016/J.ENG.2016.01.008.

Synnefa, A., Santamouris, M., & Apostolakis, K. (2007). On the development, optical properties and thermal performance of cool colored coatings for the urban environment. *Solar Energy*. https://doi.org/10.1016/j.solener.2006.08.005.

Triantafillou, T., Karlos, K., Kefalou, K., & Argyropoulou, E. (2018). An innovative structural and energy retrofitting system for masonry walls using textile reinforced mortars combined with thermal insulation. *RILEM Bookseries*. https://doi.org/10.1007/978-94-024-1194-2_86.

Velikov, K., & Thun, G. (2013). Responsive building envelopes: Characteristics and evolving paradigms. *Design and Construction of High-Performance Homes: Building Envelopes, Renewable Energies and Integrated Practice*. Franca Trubiano Ed., London https://doi.org/10.4324/9780203721797.

Wang, X., Ding, B., Yu, J., & Wang, M. (2011). Engineering biomimetic superhydrophobic surfaces of electrospun nanomaterials. *Nano Today*. https://doi.org/10.1016/j.nantod.2011.08.004.

Xing, Y., Hewitt, N., & Griffiths, P. (2011). Zero carbon buildings refurbishment – A hierarchical pathway. *Renewable and Sustainable Energy Reviews, 15*(6), 3229–3236. https://doi.org/10.1016/j.rser.2011.04.020.

Zhang, Y., Vassilopoulos, A. P., & Keller, T. (2010). Effects of low and high temperatures on tensile behavior of adhesively-bonded GFRP joints. *Composite Structures, 92*(7), 1631–1639. https://doi.org/10.1016/j.compstruct.2009.11.028.

Degree thesis

Pulsoni, G. (2016). *The use of nano-fibers in order to improve the long-term thermal and mechanical performance of polyurethane foams*. In F. Stazi, F. Tittarelli, & G. Chiappini (Eds.). Supervisors: unpublished degree thesis.

Roccheggiani. (2012). *Studio sperimentale su una finitura interna contenente pcm con progettazione e realizzazione di un mock-up: Monitoraggio invernale e calibrazione del modello analitico*. In P. Munafò, F. Stazi, & C. Di Perna (Eds.). Supervisors: unpublished degree thesis.

CHAPTER 3

A Factor to Consider When Comparing Components: The Occupants' Behavior

3.1 BACKGROUND: IMPORTANCE OF USERS' INCLUSION IN ENERGY SAVING STRATEGIES

In energy efficient buildings, the mode of use of the internal environment is a central issue. Indeed, in highly performing constructions, and even more in superefficient net zero energy buildings, the presence of overinsulated and airtight envelopes determines that the mode of use of the internal environment has a considerable impact in the final performance (Stazi & Naspi, 2018). Window openings, use of air conditioning, and the choice of the internal set point are only some of the possible controls operated by the occupants that may greatly vary from one household member to another, to such an extent that the same apartment can double its energy consumptions.

The discrepancies between predicted and real energy consumptions could reach about 30% in standard buildings (Parker et al., 2012) increasing to 50% in nearly zero energy buildings (Poirazis, Blomsterberg, & Wall, 2008).

Therefore, the prevision of final energy consumptions in the design phase could fail if the human component was not integrated and correctly represented. A more realistic model of occupants' behavior patterns will narrow the gap between predicted and actual energy use and will allow making more accurate previsions on the final efficiency of energy saving measures.

The accurate prevision of the operations on building devices, for example shading and window opening/closing, is of fundamental importance, especially in building components comparisons. Indeed, such actions modify the balance between the internal gains (due to the greenhouse effect, occupants, equipment) and the thermal loads coming through the opaque envelope.

If high window surfaces are unshaded, they will admit the incoming short wave radiation and will be instead less transparent with respect to the much longer wave radiation from the warming interior. The system will perform as a more conservative core-dominated building, thus characterized by high internal heat gains and, as a consequence, the components directly adjacent to the internal environment will have an important contribution on the energy balance. Massive heat storing internal finishing materials would be thus recommended. An active use of the windows with frequent opening actions will admit natural ventilation thus reducing the internal heat gains and nullifying the benefices of internal massive heat storing finishings with respect to lightweight ones. The same observation could be drawn for the active (or automatized) use of shadings that could reduce the convenience of adopting smart glazing systems with respect to traditional clear glasses. Moreover, the users' acceptance of low comfort levels for economic savings on energy consumptions, could make the additional cost for super-efficient envelopes unjustified.

In summary, the accurate prevision of energy efficiency of a building component requires the great variability of thermal loads due to occupants interactions with building controls to be systematically explored.

For the main aim of this book, the actions on windows (opening/closing and shading) seem to be of peculiar importance. The window shading action is not deepened since for the experimented case studies (a school and a university building), as commonly in high dimension buildings, internal louvers rather than external shading are present thus limiting the exclusion of the incoming heat wave and reducing the relevance of this action from an energy point of view. The two data sheets are consequently focused only on window-opening actions.

3.2 THE CIRCLE OF OCCUPANTS' ACTION

The prevision of occupants' behavior is very complex.

The users could behave either statically, thus passively accepting whatever condition, or dynamically, being more active in adjusting the indoor condition according to their preference and thus influencing their own future actions as well. Indeed, each occupant

Advanced Building Envelope Components. https://doi.org/10.1016/B978-0-12-816921-6.00003-2

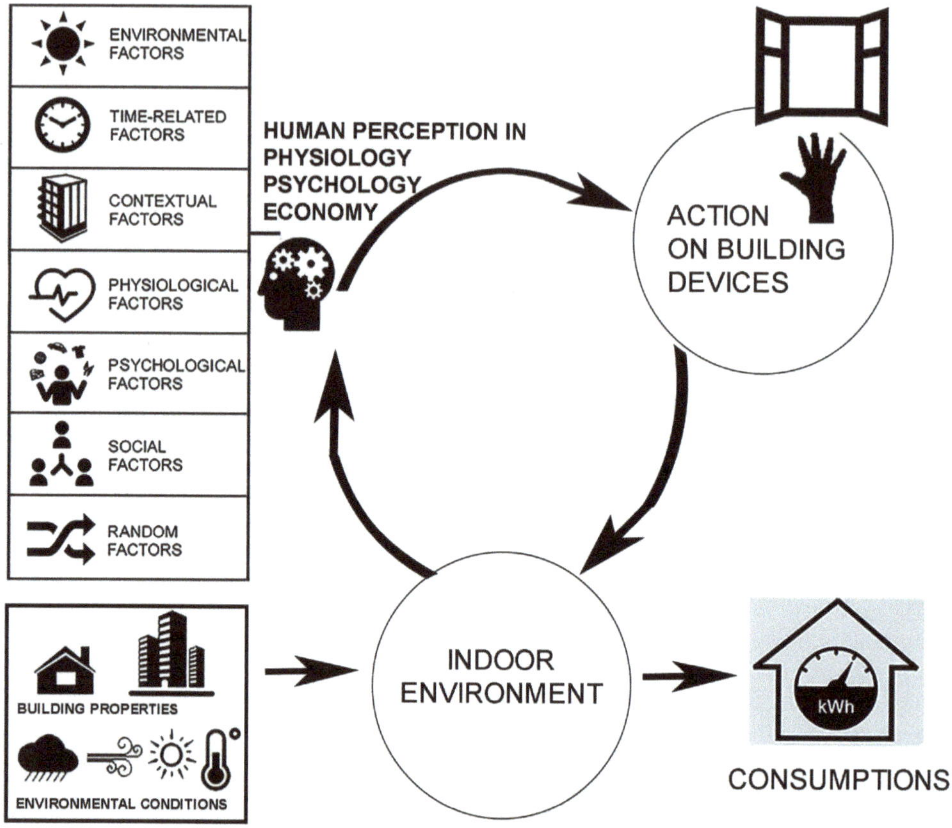

FIG. 3.1 The circle of occupants' actions.

creates a cycle (Fig. 3.1): based on his/her preferences, education, and other psychological and physiological personal factors (see Section 3.2.1), he/she decides whether the environment is comfortable or not and, in the latter case, interacts with building devices (windows, doors ...) to achieve his/her favorite sensation. Indeed, actions on building controls are governed by an adaptive principle, stating that *if a change occurs such as to produce discomfort, people react in ways that tend to restore their comfort* (Humphreys and Nicol, 1998). This suggests a feedback mechanism between the environment and the occupants' actions. The theory of adaptive comfort first conceived the building users as subjects who actively interact with the environment to modify it and adapt it to their personal needs.

As for all decision-making processes, an initial stimulus (such as a discomfort condition) is the first step; occupants perceive it and traduce this information into a response, interacting with available means (see Section 3.2.2). All the human actions on the building devices influence the indoor environmental conditions and the consequent energy consumptions and give a new starting point for the cycle continuity. Therefore, an accurate prevision of final consumptions could not be based on fixed rules regarding the users' presence and building devices position (open, closed). Nevertheless, actual energy performance simulations are based on such deterministic rules. For example, the windows are considered open when the indoor temperature overcomes a certain limit (generally 24–26°C).

Forecasting the occupants' actions is still a challenge because the complex nature of human behavior needs stochastic and complex representations. It is important to use models that accurately interpret their preferences on environment (also considering the various factors and constraints) and to reproduce analytically their probable actions with a direct effect on the environment (see Section 3.3).

3.A DATA SHEET: ACTION EFFECT ON THE ENVIRONMENT

The experimental data regarding the indoor temperature can be interpreted as temporal series. Time series analysis seeks to understand the underlined structure of sequential series of data that allows predictions of future trends. It is used in various applications such as financial data forecasting, trend detection in the global temperatures, and, in general, for the modeling of dynamic nonlinear systems.

Time series decomposition works by splitting a time series into three components: trend, seasonality, and random fluctuation. The trend is the long term increasing or decreasing value of dataset; the seasonality is the repeating short-term cycle in the series; the noise is the random variation in the series.

Although time series decomposition is generally used to eliminate randomness so that the pattern can be projected into the future, in this specific case the most interesting part is just that random quota because it is precisely due to the stochastic and non-repetitive interaction of the occupants with the environment.

The observed data in an apartment during a week in August (a) can be decomposed (Fig. 3.2) into an increasing trend (b) and a seasonal component (c), mainly a daily sinusoidal repetition of temperature connected to the particular climate and building feature. The previous steps have already extracted most of the data from the original time series, leaving behind only the "random" noise (d). It is characterized by sudden changes due to the presence and actions of occupants. It shows days of high variability in the early and later days of the series, for the presence of active occupants, while the noise is almost null from day 3 to 6 (for occupants absence) leading to a very similar observed trend of indoor temperatures (a) on such period.

Therefore, the occupants' actions perturb the environment with small random deviations, instead of creating a proper modification of the final trends, as for stable systems. According to Holling's theory (Holling, 1973), the stability of a dynamic system is its ability to return to an equilibrium state after a temporary disturbance. In other words, it means that the main trend of the system is unaffected by small perturbations. This is a prerequisite for forecasting the future trends, since only a stable system will allow reasonable predictions without great deviations from the real trends. Otherwise, it would be very difficult to follow an increasingly unpredictable road, influenced by every stochastic previous step. Thence, in this stable system, it is possible to make reasonable previsions on the real environment and as a consequence on real final consumptions.

3.2.1 Driving Factors

Various "driving" factors influence the action decision-making process. The discomfort sensation regarding thermal, visual, and indoor air quality aspects and the

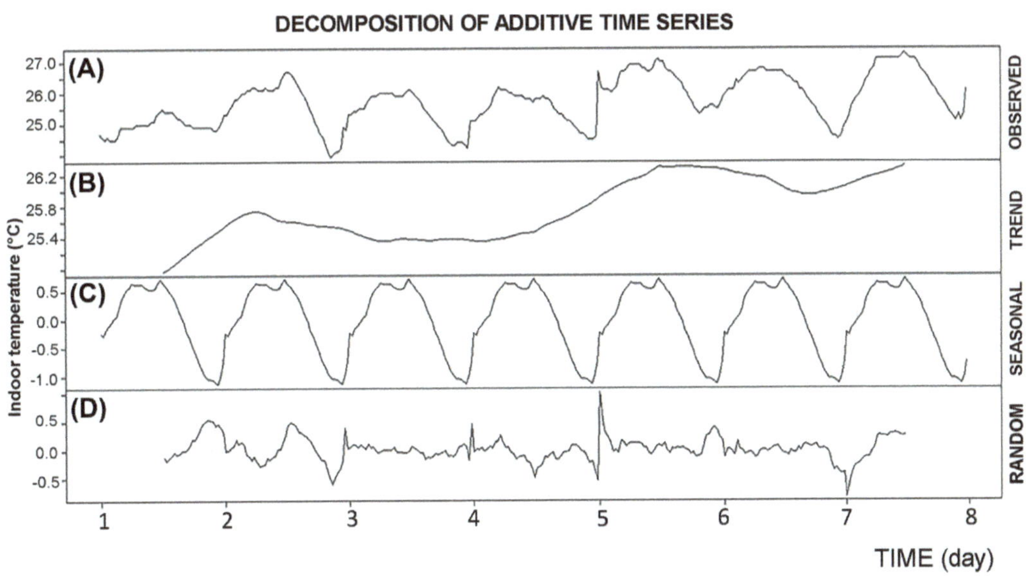

FIG. 3.2 Time series decomposition of indoor temperature of a residential building in August.

consequent decision to take an adaptive action, strictly depend on users' needs and preferences and are influenced by multiple factors. The factors that stimulate the user behavior are usually called "triggers". The following triggers are universally recognized in the literature:

- *Environmental factors*. The main factors stimulating the decision are the environmental parameters, namely temperature, humidity, air velocity, noise, illumination, odor, and CO_2 concentration. Indoor and outdoor temperatures external to the thermal comfort ranges are the stimuli for various adaptive actions such as window and blinds closing/opening, air conditioning and fan switching off/on, and heating system turning on/off. A low illuminance level invalidates visual comfort and is the main stimulus for light switching. High CO_2 concentration, strongly exceeding the IAQ comfort limits, triggers the window opening. Wind-driven rainfall is a driving factor for windows closing events.
- *Time-related factors*. The same user in the same environment could behave differently under other influencing factors, such as time of the day. For example, in office buildings the greatest use of devices (windows, lights, and shadings) is at arrival and departure times when opening and closing events are concentrated, respectively.
- *Contextual factors*. Devices accessibility affects the decision-making process. In densely occupied spaces such as offices, the occupants can take few actions for the presence of not operable devices. Moreover, the room orientation, insulation level, and heating system types are some examples of contextual drivers, indirectly influencing the actions.
- *Physiological factors*. Age and gender influence the subjective sensations, thus affecting the decision. For example, women and older subjects prefer warmer ambient than men and younger peoples.
- *Psychological factors*. Expectations affect the occupant behavior. For example, occupants of green-labeled buildings are more tolerant to discomfort, for the positive perception given by a universally recognized energy target. Experience, education, and information can have positive influences on the correct use of the controls.
- *Social and economic factors*. Working and living with other subjects, users can change their habits according to recognized group rules. For example, in schools or shared offices, the windows-opening decision is subject to others' opinion with the influence of some hierarchy (i.e., teacher consensus).

In residences, the interaction with building devices is elevated because the users feel free to manage the devices only based on their personal needs without being subject to social rules.The level and type of interaction are also strongly affected by the household economic conditions. For example, a modest economic level could lead to the reduction of the heating system set-up value and to the maintenance of the window closed for cost saving with consequent thermal discomfort and high relative humidity regimes.

3.2.2 Action Types and Priorities

Once the action decision is taken, the occupants can operate on three different levels of adaptation:

1. *Behavioral adjustment*. It regards physical changes, acting either on personal level (such as modifying the clothing or activity) or on building devices.
2. *Psychological adaptation*. It is about emotional adjustments (e.g., ignoring or tolerating discomfort).
3. *Physiological adaptation*. It means an acclimatization, intended as body changes (as increased metabolic heat production) that allow the reduction in strain induced by an uncomfortable exposure.

Among the possible adjustments, *physical action on building controls* is mediated by the other two responses, psychological and physiological. Moreover, it is the most relevant within building components optimization frameworks as it directly influences the indoor environment and consequently the energy consumptions. Thence, the literature research interest was mainly concentrated in representing and forecasting them.

The adaptive actions on building devices are several, regarding windows opening and closing, light switching, shadings and blinds use, thermostat set-up modification, air conditioning switching, fans use, and doors opening and closing.

Each adaptive action is a specific response to an environmental stimulus; inversely, the same environmental parameter could trigger multiple actions. For example, an increase of external temperature could trigger the window openings, the use of air conditioning, the fan switching on, or the use of shadings/blinds. The identification of action order and priorities is complex as boundary conditions and specific adaptive actions vary among each case study depending by the specific feature of the building. Very recent research studies are just focused on multimodal discomfort to survey what are the adaptations more likely to be adopted.

When the occupants have different choices of adaptation, *in public spaces such as offices*, they are more likely to make first personal adjustments (such as changing the clothing level) and then they interact with the most personal and nearest devices. Moreover, in such buildings the thermostat set-up is generally not modifiable by simple users. In offices, since the users commonly have tasks requiring visual engagement, the visual adaptations (for example, blind closing against glare) would have priority over thermal stimulated decisions.

In *residential buildings*, the occupants' behavior is more devoted to save energy, so hot sensations determine the decreasing of the set-point instead of windows opening, while cold sensations cause an adjustment of personal clothing level instead of the increase of the system set-point.

In naturally ventilated buildings, the interaction with windows is of primary importance.

It is worth noticing that, also in the absence of discomfort, the occupants could decide to act on a device to achieve better environmental levels based on their visual and thermal preferences rather than on the emerging need to restore their comfort. In such condition, given the diversity of personal preferences about different domains (i.e., thermal and visual), the type of decision chosen by users could be more fragmented.

3.3 BEHAVIORAL MODELS

The adaptive comfort models have recognized for the first time the fundamental importance of occupants and the necessity of considering their behavior in the design process. However, they are limited to evaluating whether or not comfort conditions are matched without giving any previsions on actions they will probably take to adjust the environment, interacting with building systems and devices. To that aim, behavioral models have been introduced.

In the last decade, various behavioral models have been proposed to consider the adaptive behaviors of the occupants and to integrate them in energy simulations. As they must rely on realistic base rather than on unfounded assumptions, all the models are developed upon observational data.

3.3.1 Types of Behavioral Models

The most simple and less accurate *deterministic models* are based on fixed and static rules, assuming that one action occurs with certainty, when a predetermined threshold value is exceeded.

More accurate and precise stochastic models have recognized the stochastic character of user behavior. Such models allow inferring a relation between the probability of one action occurrence and a series of driving factors (thermophysical parameters). Most of them are based on regression formulas, related to the identification of the users' probability (intended as a group of people) to perform a given action in relation to a stimulus, as the external or internal temperature. The formulas adopted for the relation could be of linear, logistic, or Weibull type.

In classical linear models if x was a driving variable, we would assume

$$p(x_1 \ldots, x_p) = \beta_0 + \beta_1 x_1 + \ldots. + \beta_p x_p \tag{3.1}$$

In these models, it would be possible for p to take values outside the interval [0, 1], thence when the final output is dichotomous ($0 = $ no action; $1 = $ action done), the logistic or Weibull ones are preferred.

In the logistic models, we use a suitable transformation logit (p) = ln (p(x)/(1-p(x))), and the logistic distribution will be defined as follows:

$$p(x_1 \ldots, x_p) = \frac{\exp(\beta_0 + \beta_1 x_1 + \ldots. + \beta_p x_p)}{1 + \exp(\beta_0 + \beta_1 x_1 + \ldots. + \beta_p x_p)} \tag{3.2}$$

The same research group of Nicol, that first recognized the centrality of human behavior and formulated the adaptive comfort models, in 2001 proposed (on the base of several experimental data) the first probability distribution for the prediction of windows opening as logit functions of outdoor temperature (Nicol, 2001). Rijal and Humphreys subsequently published a more refined model (Rijal et al., 2007) that, after having verified if the actual operative temperature falls without the adaptive comfort band, identified through the adaptive comfort formulas (Fig. 3.3 line 8), calculates the probability of window being open from outdoor and operative temperatures using a logit function derived by surveyed data (Fig. 3.3 lines 9 and 10).

Very interesting within this framework is also the work done by Haldi and Robinson (Haldi & Robinson, 2009) that verified several modeling approaches, including a logistic probability distribution.

Although logistic models are simply observation-based statistical correlations (without incorporating absolute threshold characteristics), *the Weibull model* encompasses both the threshold and stochastic characteristics of an occupant's response to an environmental stimulus. In this model, a triggering environmental parameter under which the probability of action is 0 must be set.

Humphreys' algorithm

No.	Window algorithm parameter	Symbol	Sample	Derivation or source
1	Outdoor air temperature	T_{out}	1 per hour	Interpolated from climate file (hourly data in file)
2	Daily mean outdoor air temperature	T_{odm}	1 per day	Calculated from 24 hourly data points per day
3	Running mean outdoor temperature (CEN)	T_{rm}	1 per day	$T_{rm}(init)=(1-\alpha)(T_{odm-1}+\alpha T_{odm-2}+\alpha^2 T_{odm-3}...)$ initial value calculated from previous 20 days daily mean, then $T_{rm}=(1-\alpha)T_{odm-1}+\alpha T_{rm-1}$
4	Running mean response to T_{out}	α	Const	Default α=0.8 (0.01 to 0.99 allowed range)
5	Comfort temperature	T_{comf}	1 per day	If $T_{rm}>10$, $T_{comf}=0.33\ T_{rm}+18.8$ (CEN Standard[(1)]) If $T_{rm}\leq 10$, $T_{comf}=0.09\ T_{rm}+22.6$ (CIBSE Guide A[(2)])
6	Indoor air temperature	T_{ai}	1 per hour	Available at each timestep (variable)
7	Indoor operative temperature	T_{op}	1 per hour	Available at each timestep (50% mrt, 50% T_{ai})
8	Comfort	Comf	1 per hour	Comf= "yes" if abs(T_{op}- T_{comf}) ≤2K Comf="hot" if (T_{op}- T_{comf}) >2K Comf="cold" if (T_{op}- T_{comf}) <-2K
9	Logit function	Func	1 per hour	Func=logit(P_w)=0.171 T_{op}+0.166 T_{out}-6.43
10	Probability function for window open	P_w	1 per hour	P_w=exp(Func)/(1+exp(Func))
11	Random number between 0 and 1	R_n	1 per hour	Generate from Fortran RNG
12	Windows status (0= closed, 1=open)	iwin	1 per hour	If Comf= "hot" and window closed (w=0) then if P_w> R_n then window open (w=1) If Comf= "cold" and window open (w=1) then if R_n > P_w then window closed (w=0)

FIG. 3.3 Humphreys' algorithm for adaptive window opening.

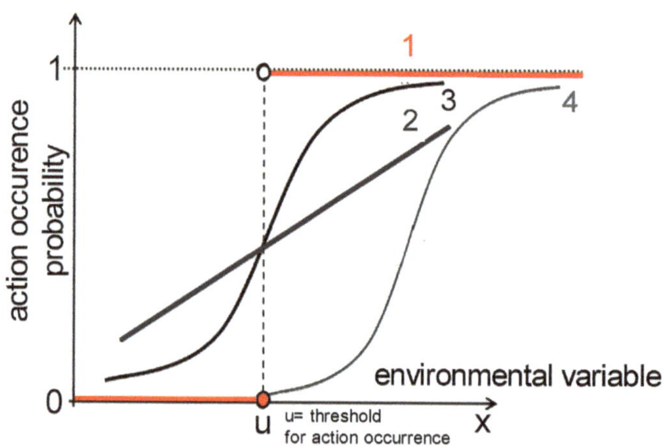

1) DETERMINISTIC MODEL $P=\begin{cases}1, & x>u\\ 0, & x\leq u\end{cases}$

2) PROBABILISTIC_ LINEAR $P(x_1...,x_p)=\beta_0+\beta_1x_1+....+\beta_px_p$

3) PROBABILISTIC_ LOGISTIC $P(x_1...,x_p)=\dfrac{\exp(\beta_0+\beta_1x_1+...+\beta_px_p)}{1+\exp(\beta_0+\beta_1x_1+...+\beta_px_p)}$

4) PROBABILISTIC_ WEIBULL $P(x)=\begin{cases}1-e^{-\left(\frac{x-u}{l}\right)^k\frac{\Delta t}{t}} & \text{if } x>u\\ 0 & \text{if } x\leq u\end{cases}$

FIG. 3.4 Graphical and analytical forms of the main behavioral models, deterministic and probabilistic.

Wang proposed a probabilistic formula with a Weibull distribution (Wang et al., 2016a) that could be used for several adjustment actions:

$$P(x_t)=\begin{cases}1-e^{-\left(\frac{x_t-u}{l}\right)^k\frac{\Delta t}{t}} & if\ x_t>u\\ 0 & if\ x_t\leq u\end{cases} \tag{3.3}$$

Fig. 3.4 reports the forms of the main behavioral models. In data sheet 3.B both logistic and Weibull models are adopted to fit the data.

The widest part of the models proposed in the literature concerns the definition of actions probability, stimulated by indoor or outdoor temperature increases based on experimental measures (Fig. 3.5). It could be

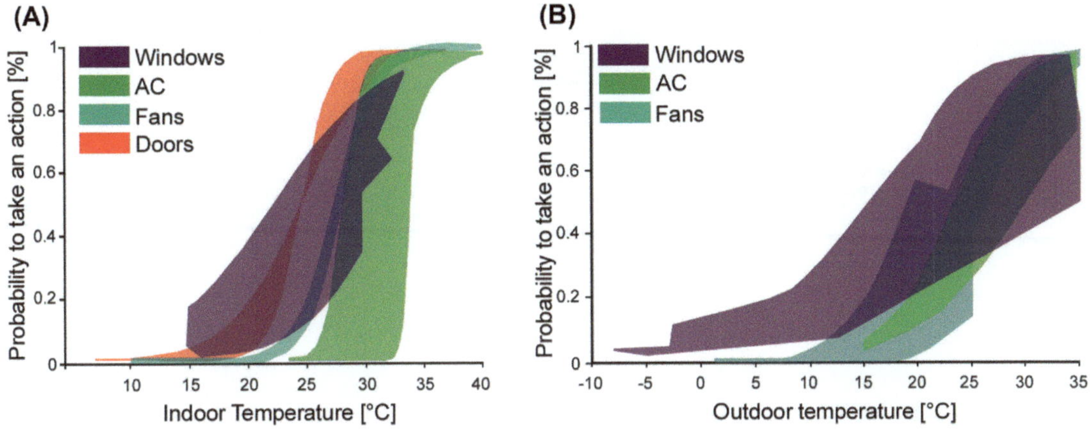

FIG. 3.5 Envelopes of the family of the regression curves existing in literature to assess the probability to take an action according to **(A)** indoor and **(B)** outdoor temperature. (From Stazi, Naspi, & D'Orazio, 2017.)

noted that even if the studies were carried out in different years and climate conditions, the trends are similar, indicating that in similar environments the reaction to a stimulus is analogous.

Evaluations regarding the statistical significance and the measure of goodness of fit are about finding relationship between the subgroups of sampled data, in an attempt to control the experiment-wise error rate. They are post hoc tests aimed at identifying if a parameter could be considered a driving factor and at determining the strength of identified correlations. For the explanation of alternative methods to evaluate the statistical significance, the reader could refer to (Haldi & Robinson, 2009).

3.3.2 Implementation in Simulations and Automatized Systems

Once the empirical model has been chosen, it can be used to forecast the user behavior and can be implemented in simulation software to enhance the performance prevision accuracy.

The probability calculation determines at each time step a probability value (between 0 and 1) of the action occurrence (i.e., window opening) depending on the value assumed by the environmental parameter that triggers the action (i.e., indoor or outdoor temperature), rather than a binary value (true = action done; false - = no action). Therefore, one simple method to assess whether the action will occur, may be to compare the probability outcome with a randomly generated number to get the deterministic signal: for example, a *P* value higher than the random number will give as output 1, that is, action done (Fig. 3.3 lines 11 and 12).

This algorithm can be included in software simulation during the run through a two-way data exchange. The engine calls the functions and provides the mandatory input to compile the algorithms. The output is sent back to the energy model, which sets the profiles and modifies the environmental conditions. This process is repeated at each time step, until the simulation end. At each time step, the device, for example a window, can turn from close ($P_{0,1}$) to open or vice versa ($P_{1,0}$), according to thermal conditions and previous status.

In the perspective of obtaining reliable stochastic results, a certain number of simulations are necessary because each simulation will produce a different history of action events and therefore different environmental temperature trends. Consequently, the main result would be a distribution of energy consumptions instead of a single value (Fig. 3.6).

3.B DATA SHEET: PASSIVE USERS VERSUS ACTIVE USERS

This data sheet aims at demonstrating that two equal rooms hosting different types of occupants, namely with different attitude towards adaptation, will have different indoor environmental conditions.

It simultaneously compares two adjacent office rooms: one occupied by persons with high attitude to interact with windows, according to their thermal sensations; the other one occupied by users suffering more passively the environmental conditions.

The two rooms are within the same building and characterized by the same geometries (Fig. 3.7).

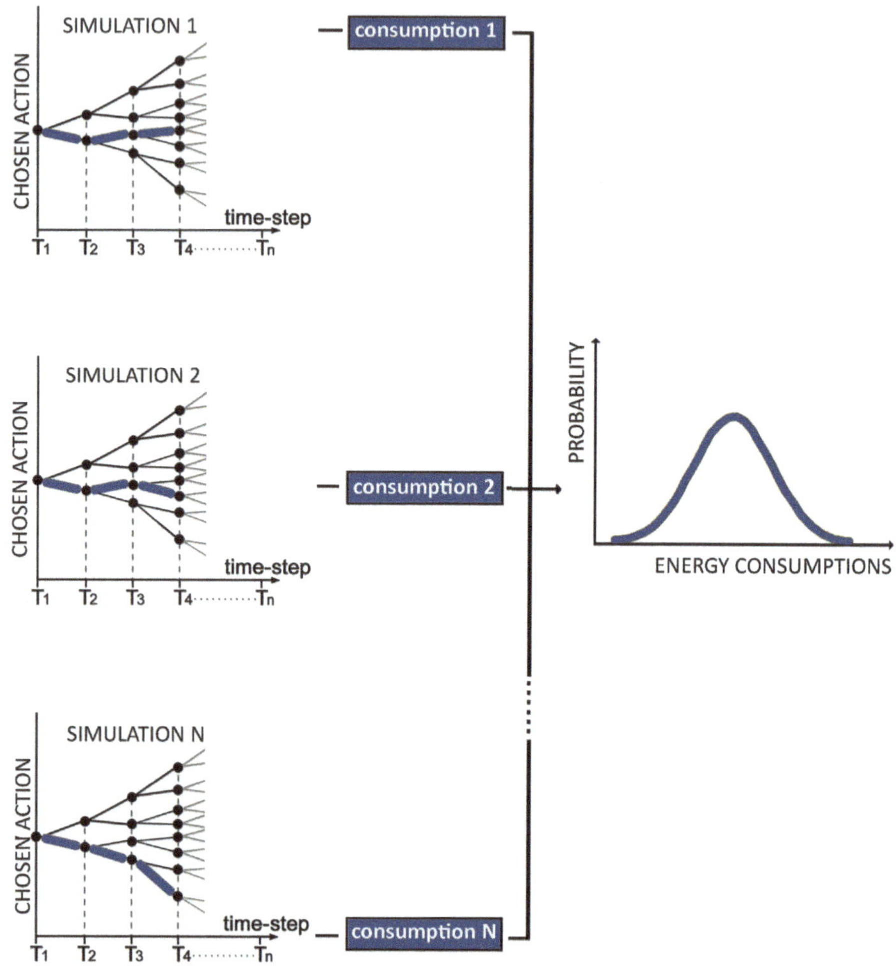

FIG. 3.6 Simulation of occupant behavior and final consumptions distribution.

FIG. 3.7 Tested rooms.

All the presented data were obtained within a Horizon 2020 founded project "NewTrends" and published in several academic papers (Naspi et al., 2018a,b; Stazi, Naspi, Bernardini, et al., 2017a).

The experimented case study was a building used for offices at Polytechnic University of Marche, in Ancona (central Italy) (see Section 4.3.1).

The evaluation of the random component (as explained in Data Sheet 3.A) on all the recorded values could give the measure of the incidence of occupant behavior on indoor environment.

The plots of the random component obtained from the indoor temperature of the not occupied rooms are drawn beside the plots of the occupied rooms for each season (Fig. 3.8). It is clear that in not occupied days the random number is always smaller than that recorded in occupied ones for the absence of people perturbing the environment.

The greatest values of random numbers (and so the greatest effect of disturb) are observed in spring when the occupants show more interaction with windows. This is because the internal temperature is increased and the external environment is not too hot yet.

The subjective perceptions on thermal sensation (Fig. 3.9) evaluated through questionnaires (PMV method) submitted to all the users revealed that the main discomfort problems are concentrated in the hottest period. Indeed, almost all the votes for the cold season fell in neutral thermal sensation (vote equal to 0).

The comparison between room 1 and 2 in winter (Fig. 3.10), when adaptive actions are not done for the presence of an external cold environment, shows that the number of days of occupation (detected by CO_2 measurement) are very similar for the two rooms. The values of indoor temperatures are slightly different in unoccupied periods, for a different switching on time of the internal heating radiators. This is clearly noticeable by superimposing the two curves (Fig. 3.11) and comparing the temperatures very near to the two radiators (as visible from the bottom image in figure).

In both rooms, the windows were always kept closed.

During the spring, the occupants' actions are instead more visible (Fig. 3.12). This is true only for room 1 where the occupants, for their habits and personal preferences, were more active on the interactions with windows (see histograms within the graphs, representing the proportion of windows open). The graph at the bottom of the combined Fig. 12 regards an enlargement of the data for a week (from first to seventh of April) by

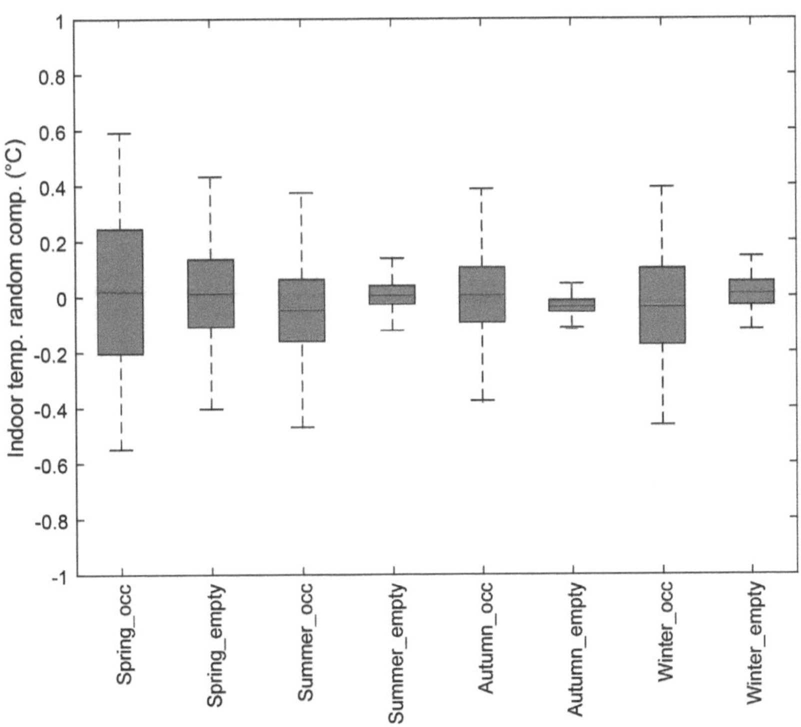

FIG. 3.8 Influence of the occupants' presence through the random component.

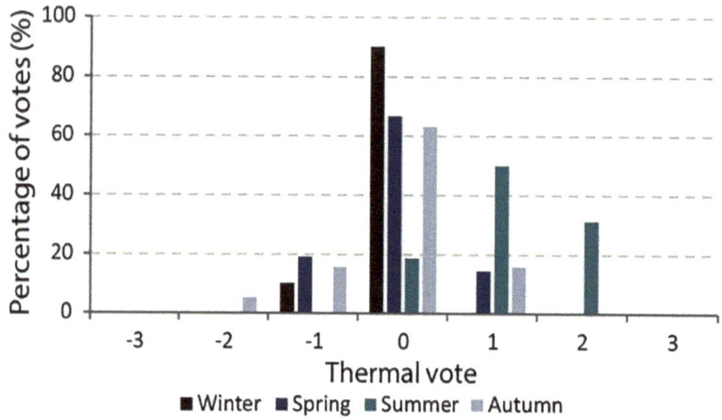

FIG. 3.9 Perceived comfort through PMV questionnaires.

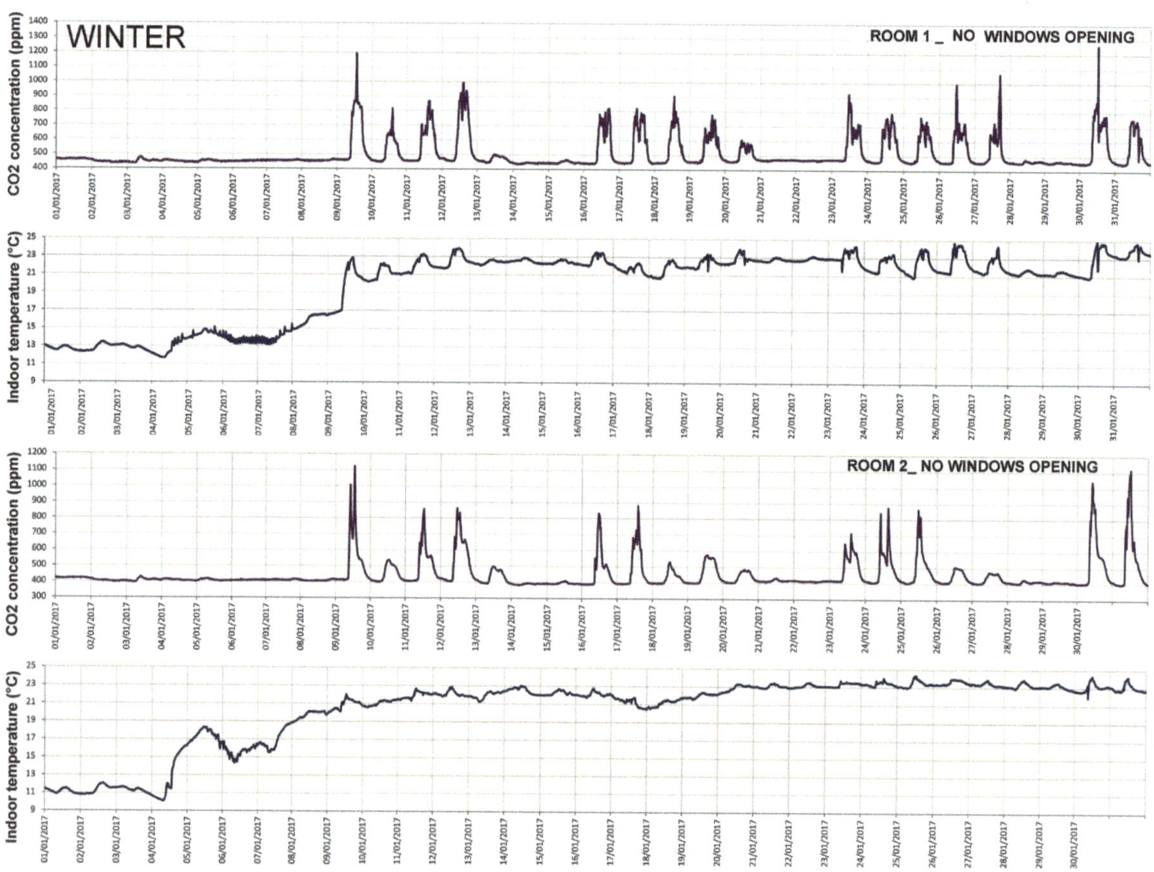

FIG. 3.10 Winter indoor temperatures and occupants' presence (through CO_2 indicator) in rooms 1 and 2.

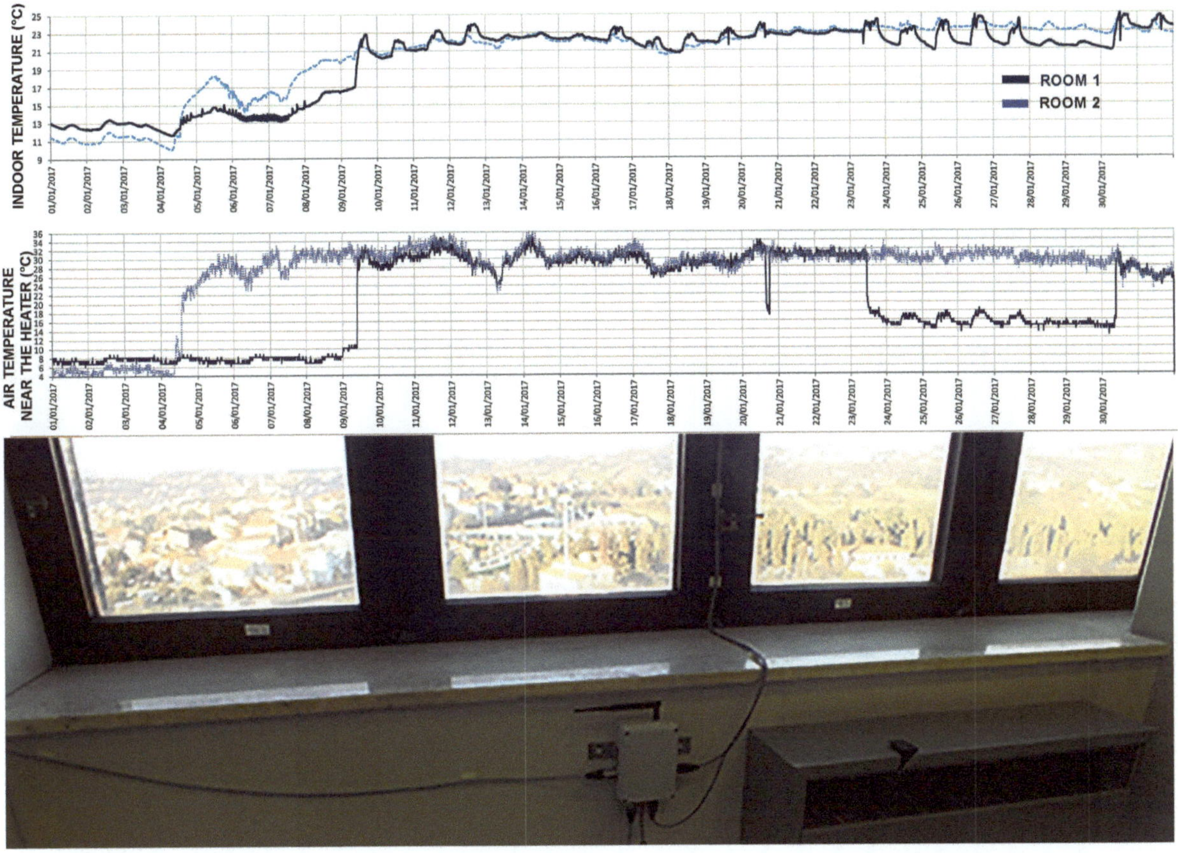

FIG. 3.11 Winter indoor temperatures and temperatures recorded near the radiators, values superimposed for rooms 1 and 2.

superimposing the data of the two rooms. Moreover, the window openings occurred for room 1 in the selected days are reported with histograms within the same graph, also reporting on the secondary y axis the quote of window openings. The openings determined a reduction of indoor temperatures (darker curve) of about 1°C and a more fluctuating trend.

Subsequently, the frequency of actions in relation to the time of day was evaluated in relation to time-related factors. In particular, the coincidence of the actions with occupants' arrival or departure times was investigated through statistical correlation estimators. The results showed that the specific users mainly open the windows at the arrival, leaving the window status unchanged until the departure time, when the greatest part of closing actions was concentrated. Moreover, the opening action eventuality at the arrival was strongly influenced both by indoor and outdoor temperatures.

Finally, the observational data were traduced in probabilistic models using both indoor and outdoor temperatures.

A logistic regression model was first adopted. The resulting models for the two rooms are reported in Fig. 3.13 for the arrival time and Fig. 3.14 for the intermediate periods.

Moreover, a Weibull distribution proposed by Wang et al. (Wang et al., 2016) was adopted to fit the data.

The parameters u, l, and k are constant values that quantify the way the occupants react to a certain environmental discomfort. The parameter u is the limit above or below which users' reaction to discomfort starts. The coefficient l is the "scale," namely the linear effect of the trigger, and k is the "shape," representing the exponent for the effect of the environmental stimulus. The temporal frequency is provided by the parameters Δt and t. The first is a discrete time step in the measurement and/or

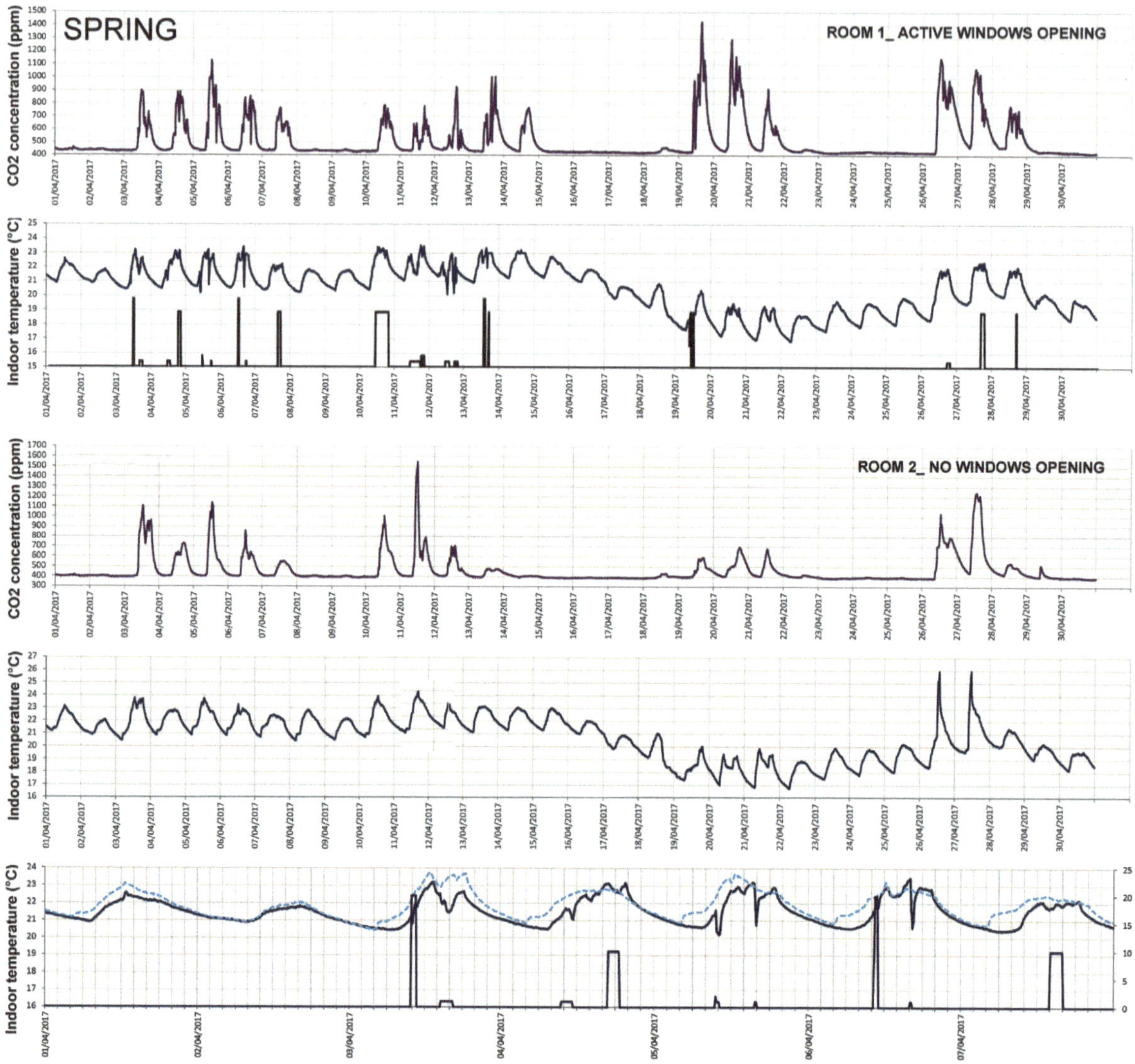

FIG. 3.12 Spring indoor temperatures and occupants' presence (through CO_2 indicator) in rooms 1 and 2.

simulation (set at 10 min) and the second is a known time constant (here fixed at 60 min).

The resulting curves combining all the surveyed data are reported in Fig. 3.15, while the parameters values are reported in Table 3.1.

The open probability depends on both indoor and outdoor temperatures and is elevated when the values overcome 25°C. Differently, the models for the closing actions with respect to outdoor and indoor temperatures significantly differ, suggesting that the closing actions are first guided by the internal temperature.

The comparison between standard, behavioral, and real window openings is reported in Fig. 3.16. During the heating season, the tuned behavioral model and the standard approach give the same prevision, considering the window closed for almost all the time, thus both slightly underestimating the real behaviors.

During the summer season, the behavioral model provides more accurate predictions than the deterministic one, with opening actions very close to those monitored. Instead, the standard profiles show an overestimation of about 30%. The standard models based on deterministic rules would underestimate winter consumptions and overestimate the summer cooling demand.

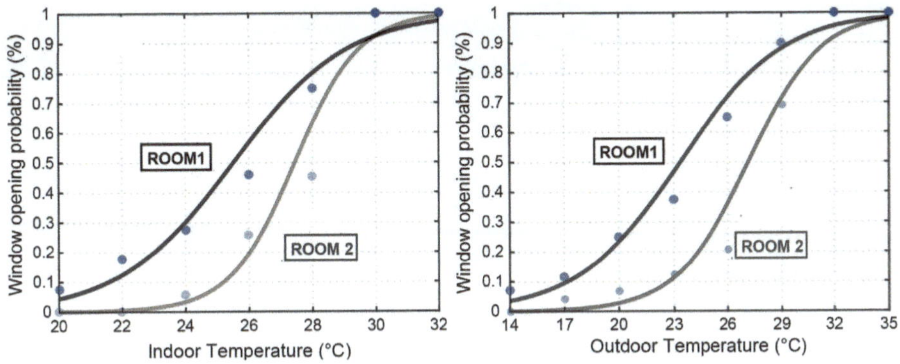

FIG. 3.13 Window-opening probability at arrival. (From Naspi et al., 2018a,b.)

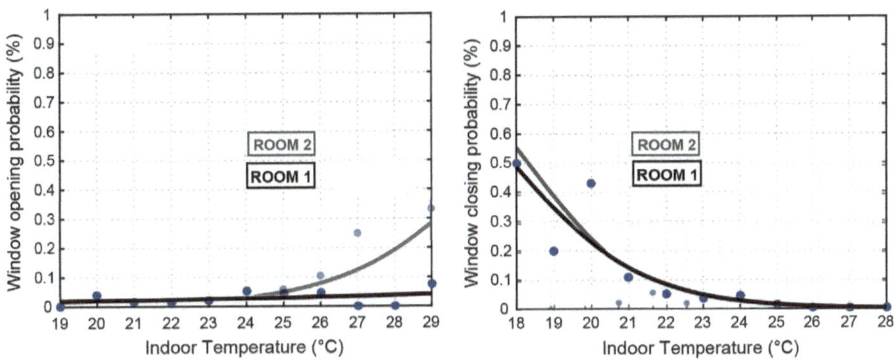

FIG. 3.14 Scarce actions (opening and closing) probabilities in intermediate periods. (From Naspi et al., 2018a,b.)

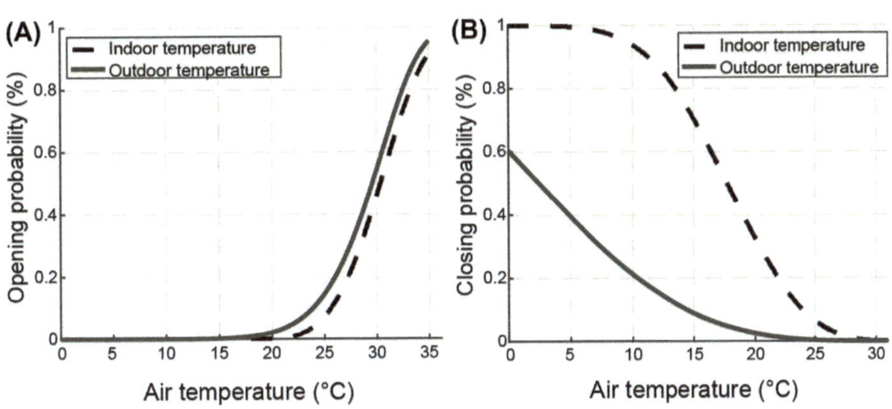

FIG. 3.15 Weibull distribution of the surveyed data. (From Naspi et al., 2018a,b.)

TABLE 3.1
Values Obtained for the Correlation Coefficients.

Correlation	u	l	k
Window open–Indoor temp.	18	8.6	3.9
Window open–Outdoor temp.	0.1	25.1	8.9
Window close–Indoor temp.	31.5	8.7	3.1
Window close–Outdoor temp.	31	18.9	3.5

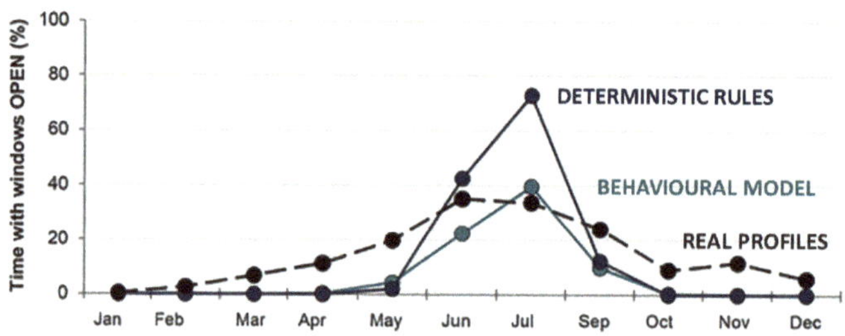

FIG. 3.16 Comparison between standard, behavioral, and real window openings, combining the data for both rooms. (From Naspi et al., 2018a,b.)

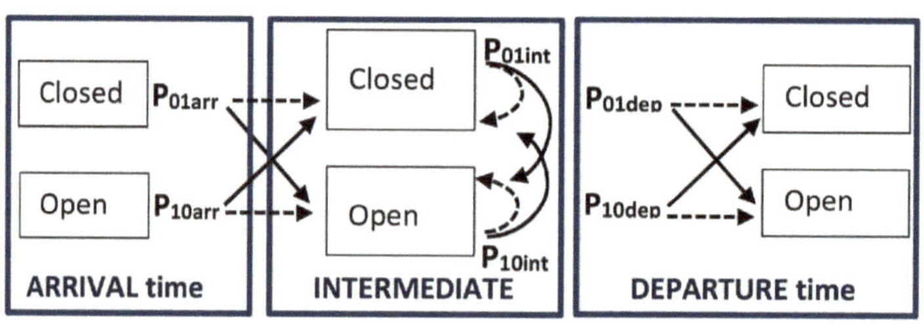

FIG. 3.17 Framework for the insertion of models within a simulation software.

Subsequently, the framework to insert the model within a simulation software has been developed (Fig. 3.17).

Actions on windows at arrival can be predicted by a specific probability (with a behavioral model suited for actions at this time). If the window is closed, the opening probability P_{01arr} will be calculated, leaving the position unchanged if the occurrence is not verified or otherwise proving the opening. If the window is open, instead, the model for closing probability P_{10arr} is recalled, and the value of probability determined for the specific temperatures and compared with the random number. This determines whether the action is done or not.

After the arrival, another submodule, launched at regular time steps, predicts the actions by relaying on two behavioral models, one for closing actions P_{10int} and one for opening actions P_{01int}. Consequently, at each time step, if the action probability is verified for a specific environmental condition (and after a comparison with the random number), then the window status changes, turning from close to open or vice versa.

When the occupants leave their offices, a third submodel predicts actions on departure. This model could also be tuned for considering the windows always closed at the arrival and always closed at the departure.

3.C DATA SHEET: USER - FREE OPENINGS VERSUS DOMOTIC CONTROL

This data sheet aims at demonstrating that proving a domotic windows control for situations in which the opening/closing decisions are committed to a group rather than to individuals could be an effective way to obtain comfort and energy saving.

The present data sheet reports the results of an experimental campaign aimed at simultaneously comparing two equal and adjacent school classrooms managed in different ways, one with automatic windows control and one freely managed by students.

The experimental study was carried out in a school building in Ancona, Central Italy (temperate climate). The monitoring details are reported in Section 4.3.2. In both classrooms, the sensors were placed at 1.1 m above the floor according to ISO 7726:2001 (ISO, 2001) for seated persons, in an almost central position. Number of occupants, adaptive actions on door and windows, and contingent events were daily noted on ad hoc spreadsheets by supervisors. To evaluate objective users' thermal comfort in the classrooms, both adaptive and PMV models were adopted. To evaluate subjective perceptions, the students and teachers in both the classrooms daily completed questionnaires. More data could be found in (Stazi, Naspi, Ulpiani, et al., 2017b).

An automatic system governing the windows opening/closing events was mounted in the two hopper windows of the motorized classroom (Fig. 3.18).

The Humphreys' adaptive control algorithm (Fig. 3.3) was chosen as driving algorithm to operate the decisions regarding opening and closing actions of the mechanized system. The monitoring probes sent data to a data logger that transmitted them to the control software developed with LabVIEW (LABVIEW 2014 by National Instruments). Every 1 minute, the adaptive control algorithm compared the values with those provided by comfort models calculation. In particular, opening/closing actions occurred when operative temperature exceeded, respectively, the upper and lower limits of the comfort band (Fig. 3.19).

The adaptive comfort temperature (T_{comf}) is that defined by Humphreys and Nicol as the optimal operative temperature and is related to the running mean of the outdoor temperature (T_{rm}). Upper and lower temperature limits for the comfort "dead band" are then defined with reference to this adaptive comfort temperature by summing or subtracting $\pm 2°C$. In some periods, the algorithm was also tested with a restricted comfort dead band (optimal comfort temperature $\pm$ 1°C); moreover, in some period also a limit value of 1250 pp for CO_2 concentration was inserted as window-opening driver.

The comparison of the classrooms during spring with still the heating system switched on (Fig. 3.20) shows that the environment is overheated in both cases. The students in free running class opened two window casings and one hopper window; the motorized systems provided the opening of the two hopper windows throughout the morning but severe indoor overheating was anyway present. The openings area was not enough to provide appropriate indoor comfort.

During nonheating season (Fig. 3.21), in the free running classroom, the adaptive actions were almost absent, for lower indoor temperatures. On the motorized classroom, some openings distributed in the morning time reduced the overheating and CO_2 concentration without threatening the comfort levels.

In summary, the interactions with windows in nonheating period are less frequent as the gradient between internal and external temperature is lower with respect to the heated season.

FIG. 3.18 Images of the windows with the motorized system.

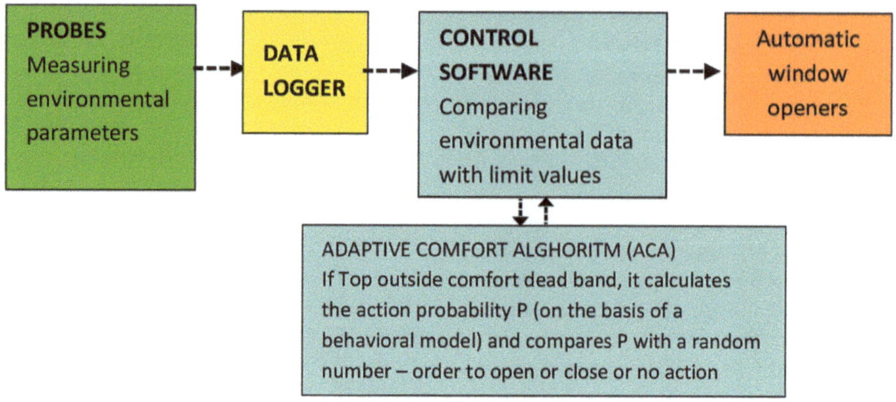

FIG. 3.19 Operating scheme of the developed system.

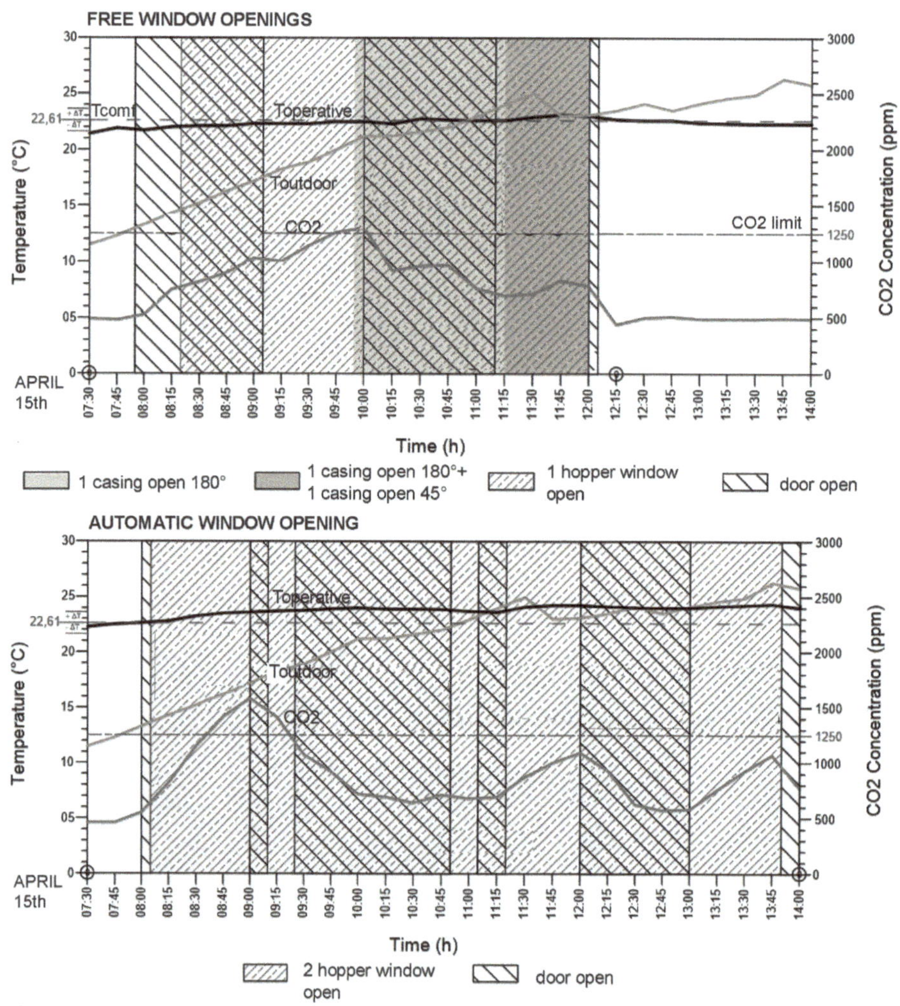

FIG. 3.20 The comparison of the two classrooms at the beginning of spring with heating system still switched on.

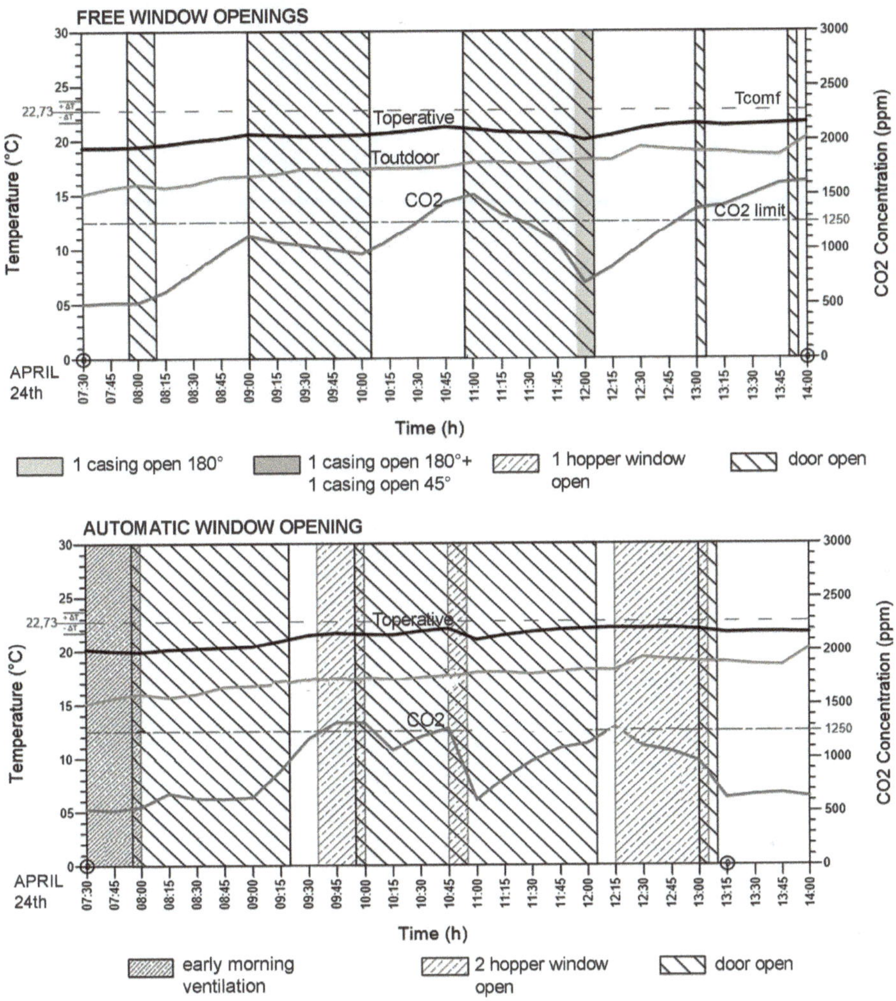

FIG. 3.21 The comparison of the two classrooms in spring.

The study of opening and closing frequency in freely managed classroom (Fig. 3.22) highlights that the actions are more likely to occur in the mid-morning interval (10.50–11.10 a.m.) and at around the beginning of new lesson hours (time intervals stressed with rectangles) as students can act on devices without interrupting lectures. The actions of opening and closing occur at the same intervals indicating that students open the windows to refresh air but suddenly reclose them.

The direct comparison of the two classrooms in spring by choosing 3 days with different outside conditions is reported in Fig. 3.23.

The 16th of April was characterized by hot temperature determining overheating, and even if the hopper windows were always open in the motorized classroom, the openings area was not sufficient to guarantee the comfort levels. The presence of a larger openable surface in the free running classroom decreed its final best performance.

The 17th and 21st of April were instead increasingly colder days. In such condition, the motorized classroom performed very better than the other one.

The summing up of the hours in which the discomfort levels exceeded the dead band limits ($T_{optimal} \pm 2°C$) for all the monitored period confirms these observations. The value is around 8% for the domotic-controlled room and about 47% for the free running one.

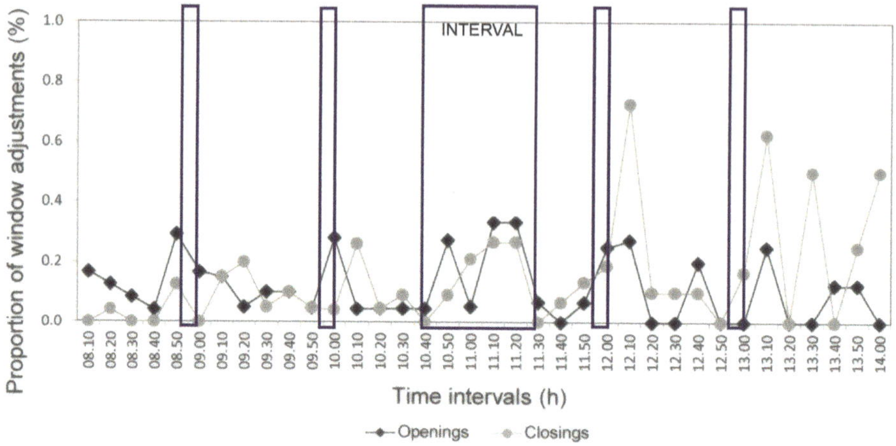

FIG. 3.22 Opening and closing events with percentage of open/closed area.

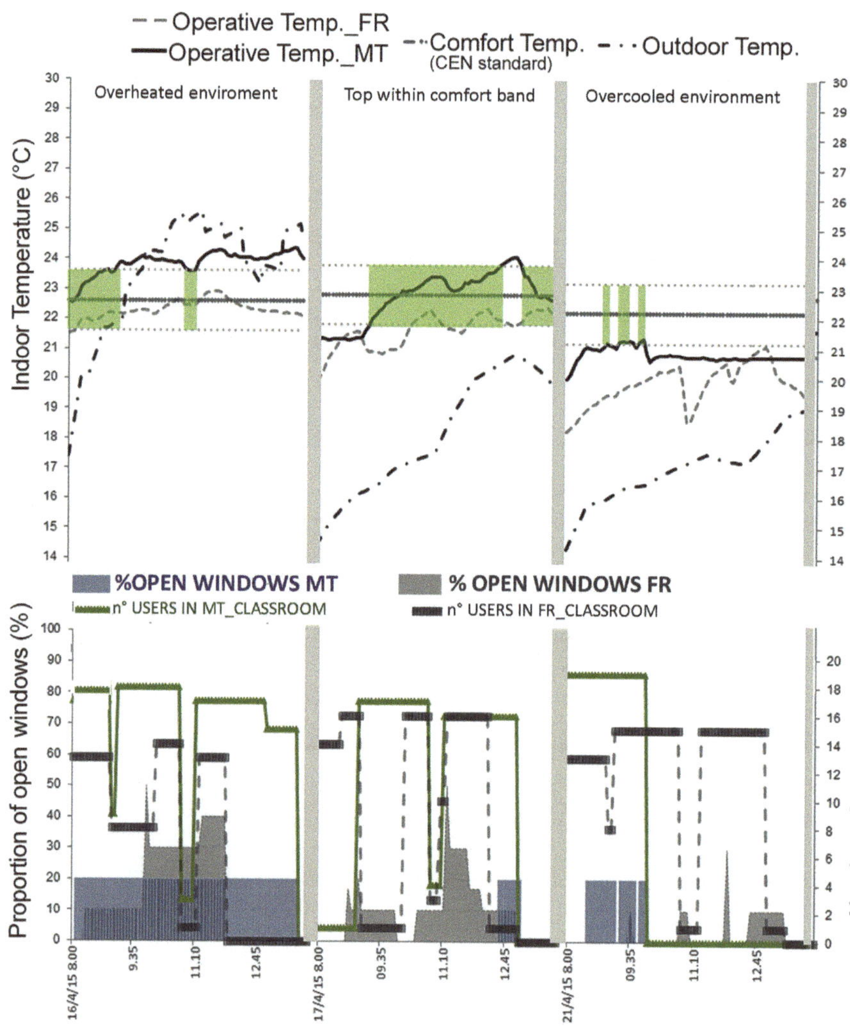

FIG. 3.23 Comparison between motorized classroom (MT) and free-running classroom (FR) in spring by choosing 3 days with different outside conditions.

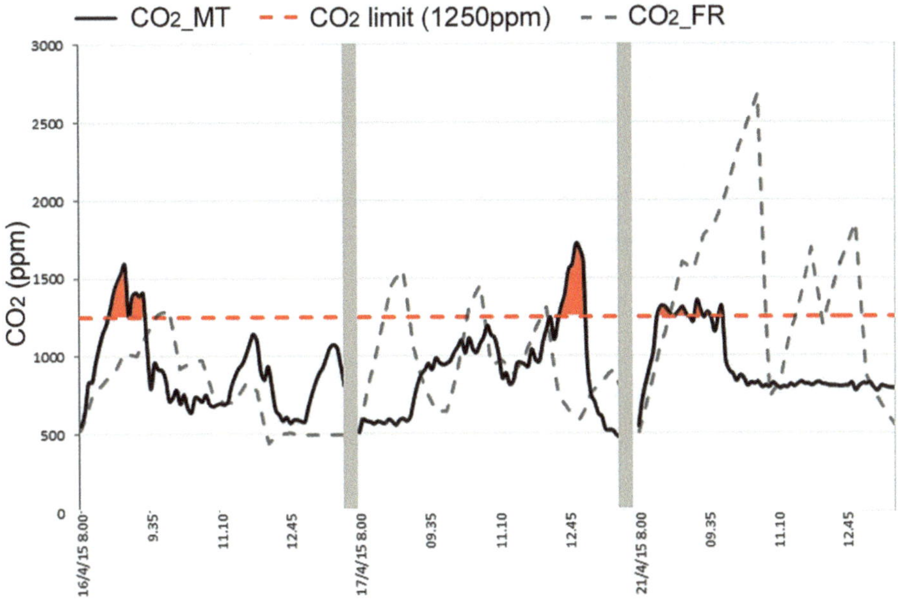

FIG. 3.24 CO_2 levels in the motorized (MT) and free running (FR) classrooms.

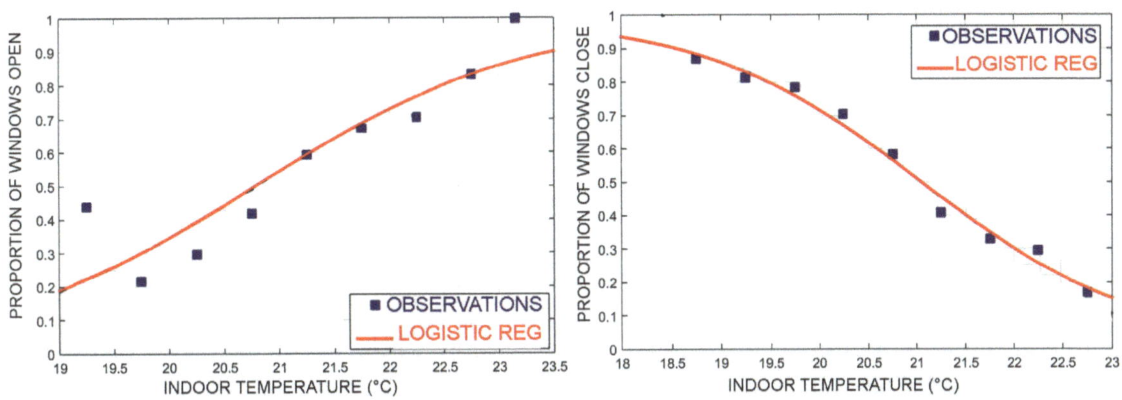

FIG. 3.25 Logistic regressions on the probability to find at least one window open and all closed with respect to indoor and outdoor temperatures.

The automated control also allowed to maintain the CO_2 concentration under critic levels (Fig. 3.24).

A logistic regression analysis regarded the evaluation of the probability to find at least one window open and all windows closed (Fig. 3.25). Indeed, in order to rely on a larger data sample, the evaluation included not only the single opening event but also the probability to find at least one window open. In the same way the window closing was translated in the probability to find all the windows closed. The relations resulted statistically significant, with R^2 indexes, respectively, of 0.81 and 0.976.

3.4 OPEN QUESTIONS AND FUTURE CHALLENGES

The inclusion of the human component within the design process requires the adoption of real cases studies to tailor the empirical variables. As a consequence, the stochastic models, strictly connected to the context in which they have been developed, may realistically forecast the occupancy-related processes and events in the selected building with its functional and organizational features, with difficult applicability to other buildings and locations (Yan et al., 2015).

However, as the generalization of the results to other buildings cannot depend on data collection and modeling task for each new building, the researchers are facing a new challenge about transportability and scalability of the results.

On one hand, some authors highlighted that the action probability driven by environmental factors was similar in different countries and buildings; on the other hand, other authors stressed that the models could be also applied on cases with boundary conditions similar to the surveyed building (e.g., settled in similar climate, usage typologies, and comparable levels of interaction with building devices).

Moreover, the extension to different contexts would be possible by tuning the coefficients of the equations through spot campaigns (instead of long-term evaluations) along the different seasons in several buildings.

The most advanced studies propose the use of immersive virtual test rooms in which several conditions can be recreated, in controlled conditions and reduced times (see Section 4.1). This multiplies the case studies and the number of users that can be involved.

Finally, a new research paradigm regards the personalized comfort models, which involves the prediction of each individual's comfort response, instead of the average response of groups. It leverages the Internet of Things, wearable sensors and machine learning to learn individuals' comfort requirements directly from the data collected in their everyday environment.

In summary, the major challenge that the researchers are facing nowadays is the development of a common framework to predict, as accurately as possible, the response of each user. They agree on the necessity to *define a standardized method to monitoring, modeling, and validating* the behavioral models to obtain comparable and shareable results. Such efforts will lead to a better understanding of users' behaviors and more accurate energy saving and thermal comfort previsions. Such previsions will be useful either for optimizing the building components design or for instructing the domotic control of the buildings devices.

REFERENCES

Haldi, F., & Robinson, D. (2009). Interactions with window openings by office occupants. *Building and Environment*. https://doi.org/10.1016/j.buildenv.2009.03.025.

Holling, C. S. (1973). Resilience and stability of ecological systems. *Annual Review of Ecology and Systematics*. https://doi.org/10.1146/annurev.es.04.110173.000245.

Humphreys, M. A., & Nicol, J. F. (1998). Understanding the adaptive approach to thermal comfort. *ASHRAE Transactions, 104*(1), 991–1004.

ISO. (2001). *EN ISO 7726:2001 Ergonomics of the thermal environment – instruments for measuring physical quantities*. Bruxelles: Comité Européen de Normalisation.

Naspi, F., Arnesano, M., Zampetti, L., Stazi, F., Revel, G. M., & D'Orazio, M. (2018a). Experimental study on occupants' interaction with windows and lights in Mediterranean offices during the non-heating season. *Building and Environment, 127*. https://doi.org/10.1016/j.buildenv.2017.11.009.

Naspi, F., Arnesano, M., Stazi, F., D'Orazio, M., & Revel, G. M. (2018b). 'Measuring occupants' behaviour for buildings' dynamic cosimulation'. *Journal of Sensors*. https://doi.org/10.1155/2018/2756542.

Nicol, J. F. (2001). Characterising occupant behavior in buildings: Towards a stochastic model of occupant use of windows, lights, blinds heaters and fans. In *7th international IBPSA conference*. https://doi.org/10.3923/jas.2013.5600.5605.

Parker, D., Mills, E., Rainer, L., Bourassa, N., & Homan, G. (2012). Accuracy of the home energy saver energy calculation methodology. In *ACEEE summer study on energy efficiency in buildings 1996* (pp. 206–222).

Poirazis, H., Blomsterberg, Å., & Wall, M. (2008). Energy simulations for glazed office buildings in Sweden. *Energy and Buildings, 40*(7), 1161–1170. https://doi.org/10.1016/j.enbuild.2007.10.011.

Rijal, H. B., Tuohy, P., Humphreys, M. A., Nicol, J. F., Samuel, A., & Clarke, J. (2007). Using results from field surveys to predict the effect of open windows on thermal comfort and energy use in buildings. *Energy and Buildings*. https://doi.org/10.1016/j.enbuild.2007.02.003.

Stazi, F., & Naspi, F. (2018). Impact of occupants' behaviour on zero-energy buildings. In *Springer Br*. Springer.

Stazi, F., Naspi, F., Bernardini, G., & D'Orazio, M. (2017a). Comparing real and predicted window use in offices. A POE-based assessment. *Energy Procedia*. https://doi.org/10.1016/j.egypro.2017.09.552.

Stazi, F., Naspi, F., & D'Orazio, M. (2017c). A literature review on driving factors and contextual events influencing occupants' behaviours in buildings. *Building and Environment, 118*. https://doi.org/10.1016/j.buildenv.2017.03.021.

Stazi, F., Naspi, F., Ulpiani, G., & Di Perna, C. (2017b). Indoor air quality and thermal comfort optimization in classrooms developing an automatic system for windows opening and closing. *Energy and Buildings*. https://doi.org/10.1016/j.enbuild.2017.01.017.

Wang, C., Yan, D., Sun, H., & Jiang, Y. (2016). A generalized probabilistic formula relating occupant behavior to environmental conditions. *Building and Environment, 95*, 53–62. https://doi.org/10.1016/j.buildenv.2015.09.004. Elsevier Ltd.

Yan, D., O'Brien, W., Hong, T., Feng, X., Burak Gunay, H., Tahmasebi, F., & Mahdavi, A. (2015). Occupant behavior modeling for building performance simulation: Current state and future challenges. *Energy and Buildings*. https://doi.org/10.1016/j.enbuild.2015.08.032.

CHAPTER 4

Experimental Methods to Compare Building Component Alternatives

4.1 ALTERNATIVE EXPERIMENTAL METHODS: POTENTIAL AND LIMITS

New trends in the field of the façade components suggest the development of suited methodologies and new facilities (Fig. 4.1).

Indeed, the new envelopes are growingly designed to interact with both indoor and outdoor environments and to involve several domains and multiple functions (i.e., thermal, visual, and air quality), integrating passive components and active control systems. Such multifunctional envelopes provide thermal insulation, control of solar radiation, controlled ventilation, and reduced condensation risk. Finally, the user behavior has a great incidence on the final efficacy of components introducing a stochastic component that needs to be evaluated.

These new directions require the integration of the traditional steady-state single-domain testing procedures with dynamic multimodal assessments. Traditional laboratory tests are nevertheless very useful to give detailed information on the effective performance of each single material and consequently guide their selection for envelope assemblies.

Studies on building components commonly are carried out through *laboratory tests on materials and assembled components*. The former regard measures on single-layered materials using well-established test equipment (e.g., Guarded hot-box) for the evaluation of conventional metrics (such as λ-value or g-value). The adoption of steady-state boundary conditions and predefined test sequences allow reaching accurate results, to control and calculate the measure errors, and to compare the results of alternative components in reliable conditions.

Test on assemblies are generally carried out on reduced scale components, such as a square meter portion of external envelope, taken as functional unit. The tests allow accounting for dynamic phenomena hardly identifiable through simplified measures, such as the flux behavior in the interfaces of stratified assemblies. This procedure could give useful information especially in comparative evaluations, but is rather limited when the aim is to establish the effective performance of a new technique under real conditions. More reliable results could be obtained in this case in *representative environments* (e.g., introducing the component in wind tunnels or climatic chambers), to reproduce the variability of boundary conditions. However, it is difficult to mimic both outdoor and indoor conditions through environment-simulating climatic chambers: the former for the complex interaction of several factors typical of the real climate, namely winds, variable temperatures and solar radiation, rain, etc.; the latter for the occupants' presence and interaction with building devices that has an undeniable and randomized impact on the environment.

On-site monitoring on real buildings are able to truthfully include all the dynamics aspects of the energy balance. Indeed, it is the sole method that is able to consider the real stochastic perturbative effect introduced by user behavior on building components and the effective external boundaries. However, this strength is also its main weakness. Indeed, this method introduces several highly variable parameters such as solar gains through windows, real life occupancy and behavioral patterns, variable set points, internal gains from electrical devices, and cross-ventilation. Each variable can hardly be isolated from the others, thus leading to misinterpretations. Moreover, it links the results to a variety of contextual factors, such as building geometry and orientation (of each component and opening), local meteorological profiles, and specific set and tuning of Heating Ventilation and Air Conditioning systems. However, this experimental method is very useful when the evaluations are of comparative types. Moreover, the results are transferable to other building typologies and climatic context with the aid of properly calibrated simulation models.

The adoption of a test room is an in-between solution, which retains the complexity of an environmental external—internal interaction without the actual uncontrollability of a real-scale fabric (Cattarin et al., 2016). Indeed, the external fabric is exposed outdoor while

Advanced Building Envelope Components. https://doi.org/10.1016/B978-0-12-816921-6.00004-4

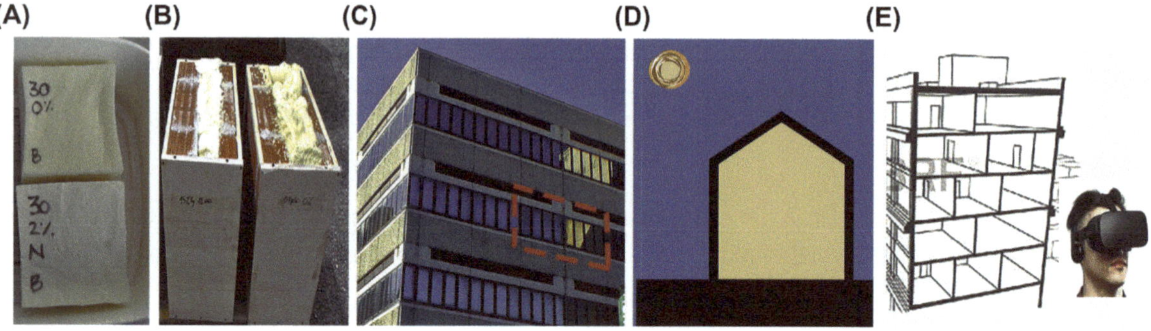

FIG. 4.1 Laboratory tests on materials **(A)** and components **(B)**; on-site monitoring **(C)**; test rooms **(D)**; virtual immersive environment **(E)**.

the internal environment mimics a realistic indoor condition through air conditioning, HVAC systems and lamps. To avoid uncontrolled and not uniformly distributed gains, the test facilities are commonly without windows and the variable heat gains due to daylight cycle through transparent surfaces (as occur in a real environment) are reproduced through active systems. Moreover, to avoid discrepancies between mock-up based results and effective performance, the possible occupants' perturbations could be introduced through the active control of the heat gains/losses.

This method should be complemented with laboratory tests for a more accurate identification of conventional metrics (such as steady state *U*-value), more difficultly achievable in facilities exposed to variable boundary conditions (Goia et al., 2017).

A recent challenge that the researchers are engaging on, is the integration of physical built environments (e.g., a test room) with *immersive virtual environments* to conduct controlled behavioral studies (Heydarian et al., 2015). In this method, the environment is virtually reproduced within a test room through head-mounted devices (HMDs) and virtual reality glasses (e.g., Oculus Rift) using desktop 3D graphics. A set of identical 3D virtual models of the real test room with the same dimensions, furniture, and similar lighting conditions has to be created. Variations in environmental conditions (i.e., illumination levels) can be easily implemented. Obviously, some measures have to be taken to provide a responsive environment with respect to the users' decisions (i.e. systems able to really perform the chosen action), as they will experience the outcome of the interaction and this will affect their following decisions.

4.2 LABORATORY TESTS

The comparative evaluations reported in this work were, in the initial phase, based on several laboratory tests regarding the thermohygrometric properties, the chemical and morphological structure of the materials, and the effect of aging on the long-term performance.

4.2.1 Thermal Conductivity Analysis

Laboratory tests for the determination of the thermal conductivity were carried out on samples of various materials by measuring the average thermal flux exchanged in steady state. The heat flow meter method described in standard EN 12667: 2001 for materials characterized by medium and high thermal resistance was selected.

The test system (Fig. 4.2) is composed of a box of thick insulating material that is closed during the execution of the test to ensure that the test environment is adiabatic with respect to the external environment.

A unidirectional constant and uniform density of heat flow rate is generated by adopting a hot and a cold plate connected to thermostatically controlled water baths, at the temperature of 30°C and 10°C, respectively. This condition is verified with acceptable accuracy in the test apparatus center; the apparatus is therefore divided into a central metering section in which measurements are taken and a surrounding guard section.

Temperature probes and flowmeters on both sides of the sample measure, respectively, the thermal gradient and the crossing thermal flux (q). The test duration should ensure the reaching of stationary conditions: this condition can be considered satisfied when the thermal flows recorded by the two flowmeters reach almost equal values. In steady-state conditions, in fact,

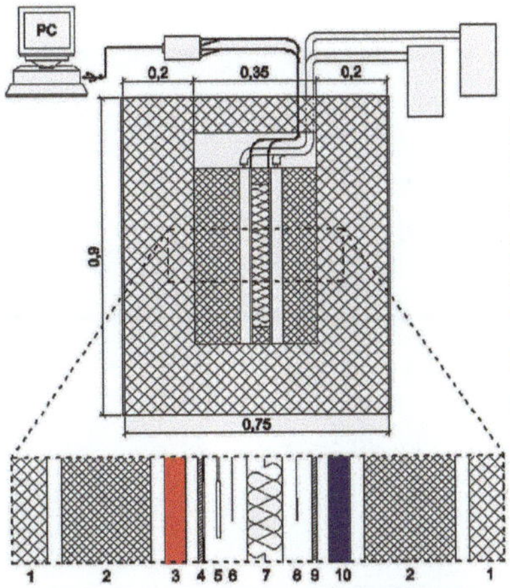

1 EXTERNAL BOX
2 INSULATION
3 HOT PLATE
4 WOODEN PLANK+RUBBER SHEET
5 FLOW METER
6 TEMPERATURE PROBES 1 AND 2
7 SAMPLE
8 TEMPERATURE PROBES 3 AND 4
9 WOODEN PLANK
10 COLD PLATE

FIG. 4.2 Heat flow meter method apparatus.

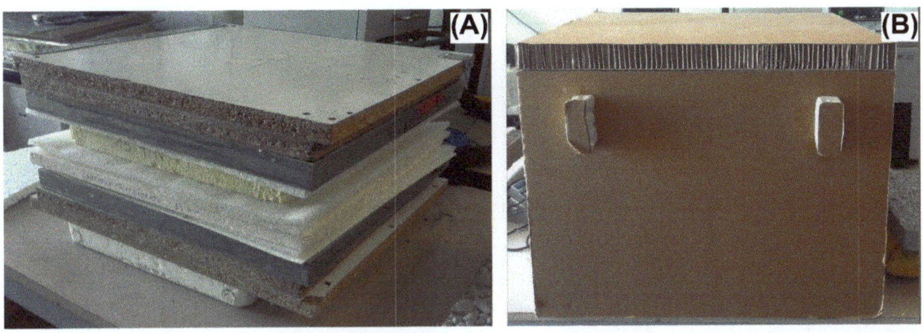

FIG. 4.3 Preparation of the layers **(A)** that subsequently will be inserted within the insulated box **(B)**.

the flow is unidirectional and constant in all the surfaces of the system. The duration of the tests resulted of 24 hours for all the samples. An initial calibration was carried out to verify the measurement error, using a panel in XPS (extruded polystyrene) with a known conductivity value.

The specimens should be homogeneous with flat parallel faces, in the form of slabs.

For the necessity to maintain the samples in vertical position within the box, a layered structure with wooden planks was developed (Figs. 4.3 and 4.4). Metal screws and plates were used to assemble the layers together.

The samples were realized with a surface of $50 \times 50\ \text{cm}^2$.

The test procedure includes the following steps: smoothing of the specimen surfaces to obtain uniform thickness; assembling the test system making all the faces adhere to avoid the creation of air layers; insertion of the system inside an insulating box, and sealing with respect to the external environment; imposition of the thermal gradient and recording the resulting flux.

With the data measured in the test, it was possible to obtain the resistance value and, therefore, the value of thermal conductivity through the following equations:

$$R = \Delta T/q;\ \ \lambda = t/R \rightarrow \lambda = (t \times q)/\Delta T \qquad (4.1)$$

where t is the thickness of the material sample (m), λ is the thermal conductivity (W/m K), R is the thermal resistance (m^2 K/W), q is the density of heat flow rate (W/m^2), and ΔT is the difference in temperature between the two surfaces of the material (K).

The test was also adapted for assemblies (Fig. 4.5A) rather than single-layered materials. The assembly was inserted within the box connected to the thermal bath (Fig. 4.5B and C) and, after the box upper closure, the thermal conductance C was determined.

4.2.2 Moisture Content

The moisture content test is applicable to porous water permeable materials and allows calculating the ratio between the sample mass before and after drying at

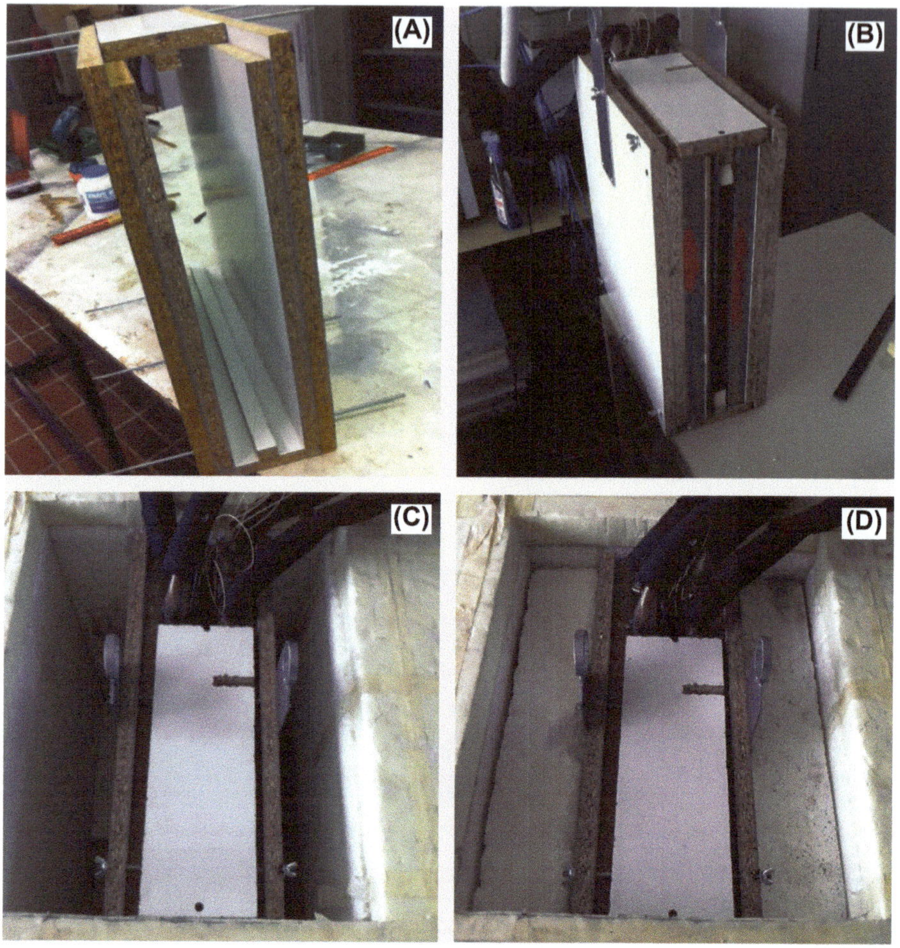

FIG. 4.4 Phases of equipment assembly: **(A)** wooden planks; **(B)** fixing with metal screws; **(C)** insertion of the component within the box; **(D)** insulation of both sides of the sample.

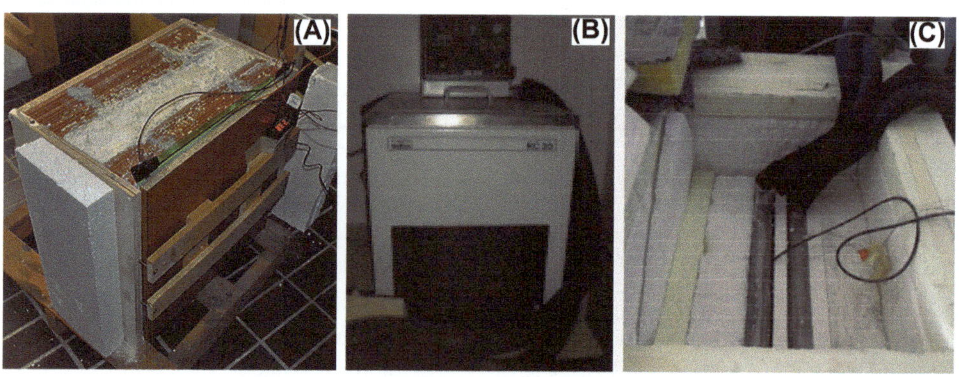

FIG. 4.5 Phases of insertion of the components within the insulated box.

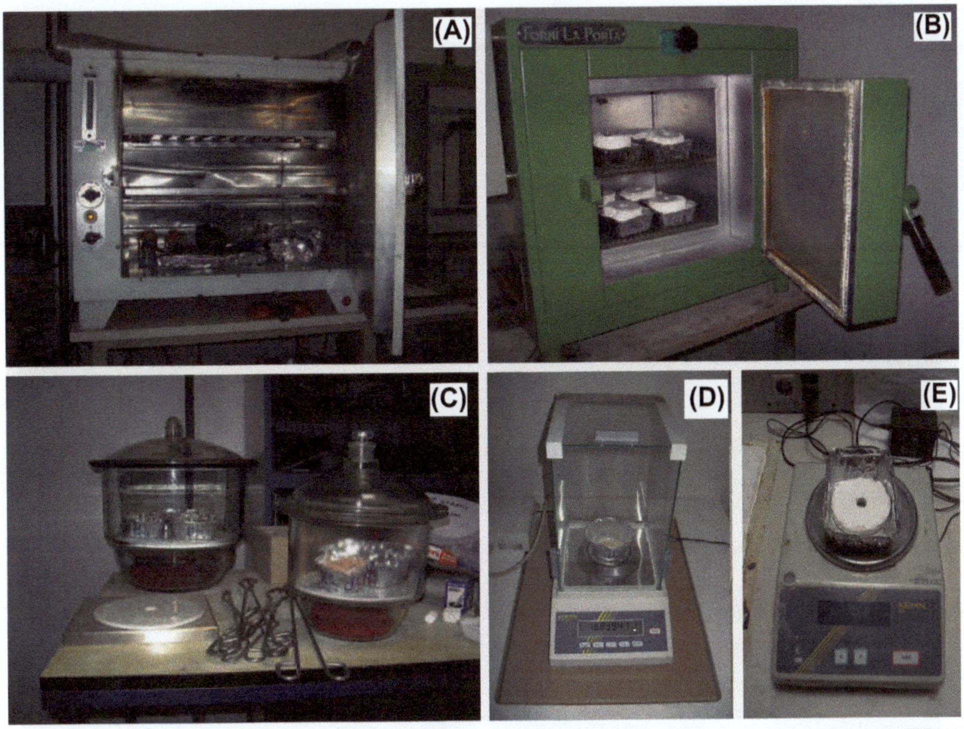

FIG. 4.6 Thermostatic oven set at 105°C **(A)** and 40°C **(B)**; desiccators **(C)**; precision balances KERN 770 **(D)** and KERN 470 **(E)** for different weighing ranges.

elevated temperature in a thermostatic oven, according to EN ISO 12570.

The test equipment (Fig. 4.6) indicated by the standard provides a ventilated oven able to maintain a specifically selected drying temperature with the indicated accuracy (e.g., 40 ± 2°C for foams and 105 ± 2°C for mineral materials) and relative humidity less than 10%, a desiccator, a scale or balance capable of weighing specimens with an uncertainty not greater than 0.1% of their mass, and tools to measure specimens dimensions.

The minimum number of samples is not specified by the standard. If extracted from large elements, they should be taken in such a way that cutting and drilling do not significantly affect the moisture content. The mass of the samples before drying should be measured immediately after the cutting operation; otherwise, the standard specifies how to store the samples in vapor-tight containers and measure the packing material weight.

The test procedure provides that the laboratory temperature during the test is kept at 23 ± 6°C.

Before drying, the specimens must be weighed with an accuracy equal to 0.1% of their mass.

The specimens have to be dried using the oven at the temperature specified earlier, up until the constant mass condition has been reached. This condition occurs when the mass variation in three successive weighing carried out at 24 h intervals is less than 0.1% of the total mass. The test specimens must be cooled (at 30–40°C) in the desiccator and weighted.

According to the reference standard, the moisture content mass by mass, u, of a test specimen should be calculated with the following equation:

$$u = (m - m_0)/m_0 \tag{4.2}$$

where m is the mass of the specimen before drying (kg), and m_0 is the mass of the dried specimen (kg).

4.2.3 Hygroscopic Sorption Properties

The hygroscopic sorption test allows evaluating the moisture content of a building material in equilibrium with air at a specific temperature and humidity. The hygroscopic sorption properties were determined according to ISO 12571:2013. This standard specifies two alternative methods: desiccator method and the climatic chamber method. The former is used hereafter.

The test equipment includes a balance KERN 470 with an accuracy of ±0.01% of the mass of the specimen; drying oven (M710 Ceramic Instruments S.r.l. was

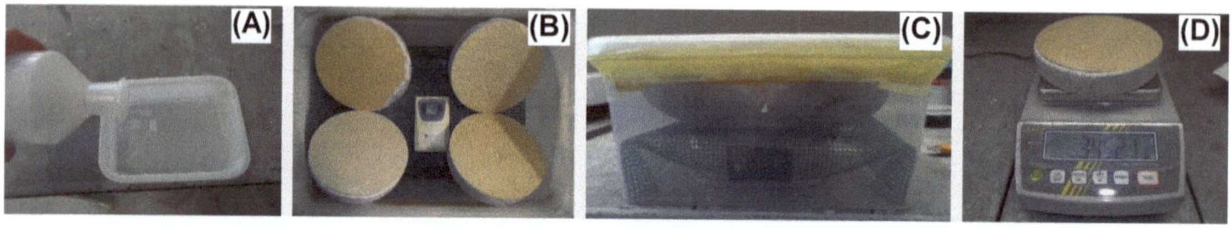

FIG. 4.7 Saturated solution **(A)**; sample positioning **(B)**; desiccator **(C)**; precision balance **(D)**.

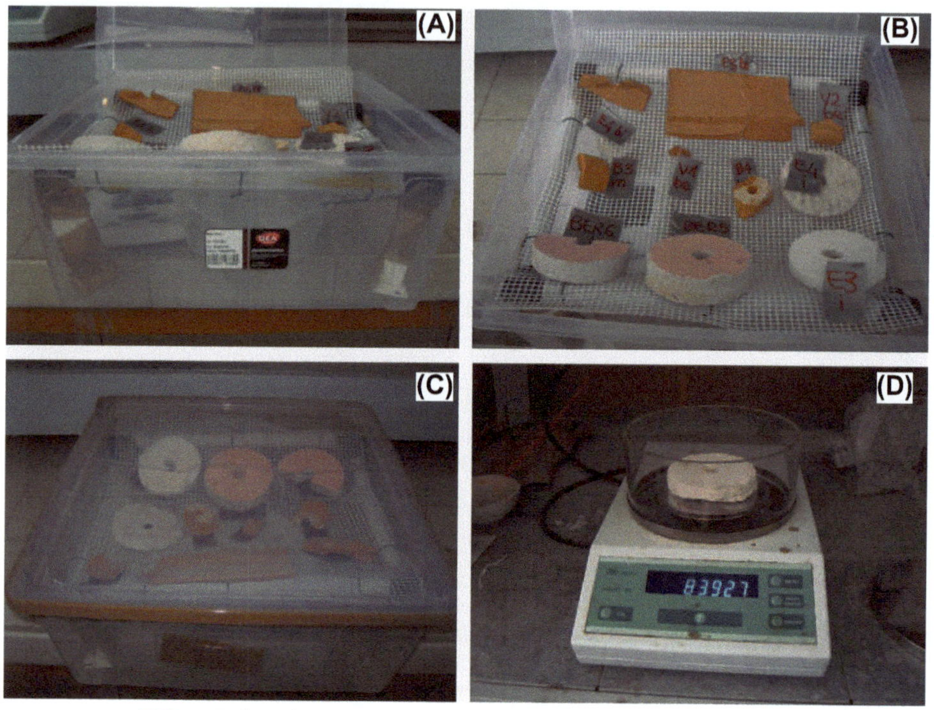

FIG. 4.8 Samples in the desiccator **(A, B, C)**; precision balance **(D)**.

selected); desiccator, capable of maintaining the relative humidity value within ±2% RH; constant temperature chamber, able to maintain a fixed temperature with ±0.5 K accuracy.

Furthermore, the following saturated aqueous solutions were used to achieve the necessary relative humidity inside the desiccator:

KOH for RH = 9%,
$MgCl_2$ for RH = 33%,
$Mg(NO_3)_2$ for RH = 54%,
$NaCl$ for RH = 75%,
KNO_3 for RH = 93%.

A plastic mesh was used to support the samples inside the desiccator (Figs. 4.7 and 4.8), thus avoiding the direct contact between samples and saturated solutions.

The plastic net with the samples, the saturated solution (depth 40 mm), and a hygrometer for measuring the humidity were all introduced within the desiccator; then, it is closed with a lid and sealed with adhesive tape.

According to the standard, a minimum of three specimens should be tested. The specimens must be representative of the product and have a mass of at least 10 g. Specimens of materials with a dry density less than 300 kg/m^3 must have a surface of at least 100 mm × 100 mm. As regards the polymeric foams and Glass Fiber Reinforced Polymer (GFRP) profiles, the samples were tested both with and without the external surface veil, since such external coating is recognized to change the material sorption properties.

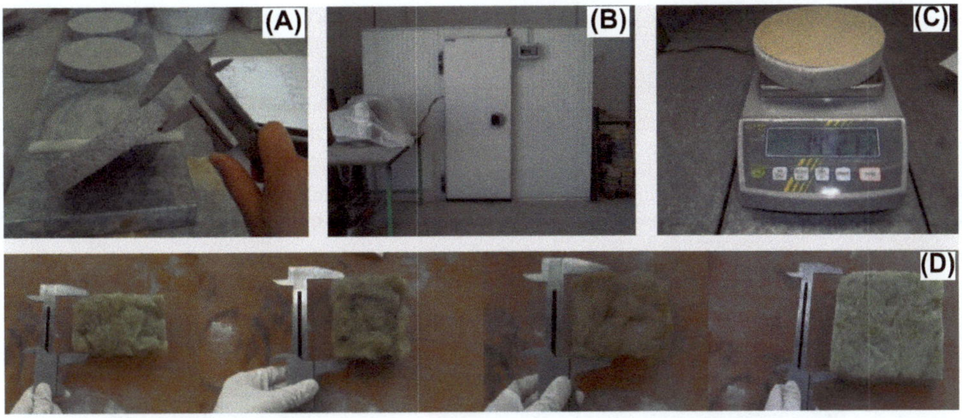

FIG. 4.9 Apparent density: phases of measure **(A)**, conditioning **(B)**, and weighing **(C)**, linear dimension measuring for various samples **(D)**.

The test procedure consists of drying samples in the drying oven at suitable temperature for the tested material (as specified in ISO 12570), until reaching a constant mass. For example, a temperature of 40°C is suitable for foams, and 110°C for mineral wood. Constant mass is reached when the change in mass between three successive weightings, performed at intervals of at least 24 hour from each other, is less than 0.1% of the total mass.

The specimens are then inserted within the climatic chamber at constant temperature ($T = 23 \pm 0.5°C$) and weighed periodically until they are in equilibrium with the environment (constant mass).

Then, the relative humidity has to be increased (replacing the salt solutions) by choosing at least four RH values evenly spaced within the range of 30%–95% relative humidity. For each RH value, the procedure of sample weighting is repeated and the equilibrium moisture content curves are drawn.

The moisture content, u (kg/kg), is calculated for each specimen by

$$u = (m - m_0)/m_0 \tag{4.3}$$

where m is the mass of the specimen at each relative humidity level (kg) and m_0 is the mass of the dried specimen (kg).

The average of the calculated values of the moisture content for the various specimens is considered at each humidity value. Then, the values are reported in a diagram as a function of the relative humidity values (hygroscopic sorption curves).

4.2.4 Apparent Density

The apparent density refers to the ratio of the mass and the volume of the test specimen. The volume is that occupied by a solid, including hollow spaces and all the surface skins formed during production. The apparent density depends on the form of the particles, the composition of the substance and the storage method. It was determined according to standard EN 1602:2013.

The test equipment indicated by the standard includes the following:

Balance able to determine the mass of the sample with an accuracy of 0.5%; equipment for determining linear dimensions with an accuracy of 0.5%; drying oven. At least five specimens should be tested with as large as possible dimensions and simple shapes for easy volume calculation. The specimens must be cut with methods that do not alter the original structure of the product.

The test procedure (Fig. 4.9) involved the following operations:

With the use of a gauge, at least six measurements were made for each dimension of each specimen and the values were averaged. The specimens were subjected to predrying in the oven at a temperature of 40°C to reduce the conditioning time. The samples were then placed in an air-conditioned chamber with stable temperature of $(23 \pm 2)°C$ and relative humidity of $(50 \pm 5)\%$ and weighted with precision scales at 24-hour intervals until constant mass was achieved.

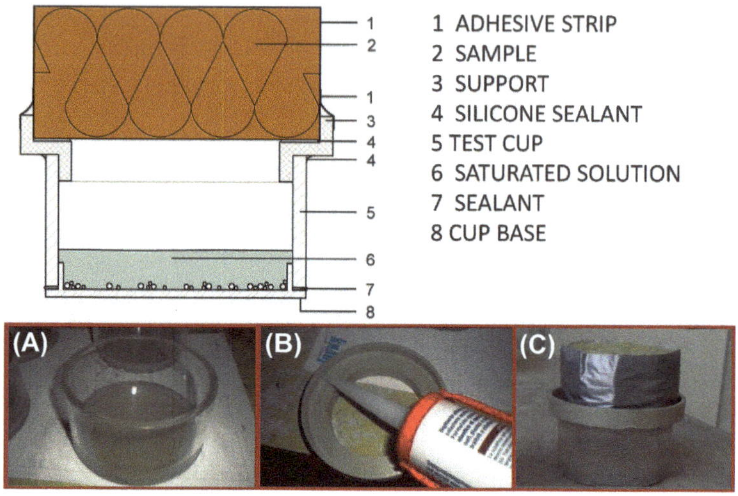

FIG. 4.10 In the top of the figure, a section of the test cup shows all the components; at the bottom, three images **(A, B, C)** show the realization method.

Average apparent density, ρ, was calculated as the ratio between average weight and volume of the samples:

$$\rho = m/V \quad (4.4)$$

where m is the mass of the specimen (kg) and V is the volume of the specimen (m^3)

4.2.5 Water Vapor Transmission

The water vapor transmission test allows determining the water vapor permeability of building materials under isothermal conditions. The test was used to gain information on the performance of materials at high humidity conditions. Wet cup method of ISO 12572: 2016 standard was adopted.

The test equipment includes the following:

Test cups (preferably circular and transparent) resistant to corrosion from the saturated solution (Fig. 4.10); analytical balance with accuracy of 0.001 g; a climatic chamber for temperature and humidity control able to maintain RH values within ±5% and temperatures within ±1 K; probes to measure temperature and humidity; sealant impermeable to water vapor.

The minimum number of samples to test is five if the specimen area is less than 0.05 m^2, otherwise three. The specimen has been cut so that the larger parallel faces are those traversed by the water vapor.

The exposed (upper and lower) area of the specimen must be at least 0.005 m^2 and the diameter must be twice its thickness.

A saturated potassium nitrate solution (KNO_3), which gives an RH = 94% at $T = 23°C$, was adopted.

The test procedure provides that the specimens are preliminary conditioned at 23 ± 5°C and 50 ± 5% RH until weight stabilization, so that three successive daily weight measures agree within a 5% error; a storage time of 6 hours is anyway necessary. The test specimen is then sealed into the open side of a test cup containing the aqueous saturated solution. The solution is placed inside the cup for a height of about 15 mm from the bottom leaving an air gap below the sample surface of 15 ± 5 mm. The assembly is then placed within a climatic chamber at RH = 50 ± 5% and $T = (23 \pm 5)°C$.

Due to the difference in partial vapor pressure between the test cup and the chamber, a vapor flow occurs through the specimen. The mass loss by the specimen is measured during time through periodic weighing within the test chamber and the steady-state rate of water vapor transmission is thus determined.

For each pair of successive weighing of the specimens, the mass change rate Δm_{12} is calculated using:

$$\Delta \dot{m}_{12} = (m_2 - m_1)/(t_2 - t_1) \quad (4.5)$$

where $\Delta \dot{m}_{12}$ is the change in mass per time for a single determination (kg/s), m_1 is the mass of the test assembly at time t_1 (kg), m_2 is the mass of the test assembly at time t_2 (kg), and t_1 and t_2 are the subsequent measurement times (s).

The representation of a graph of mass variation over time allows to calculate the slope G (kg/s) representing the water vapor flow rate through the specimen.

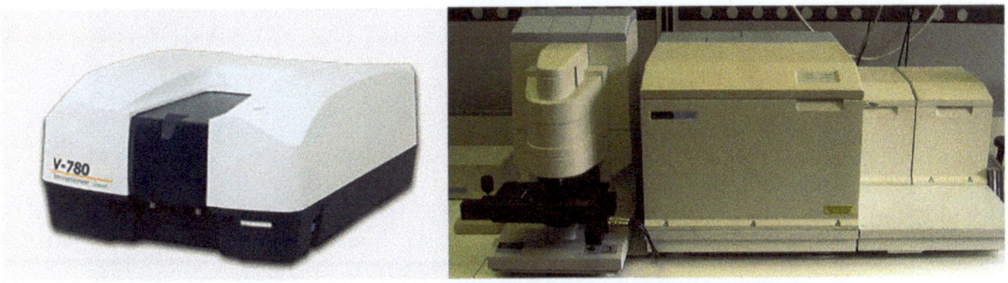

FIG. 4.11 Tools used for spectrophotometric tests.

The density of water vapor flow rate g (kg/s m^2) is calculated as follows:

$$g = G/A \quad (4.6)$$

where A is the exposed area of the specimen (m^2).

The water vapor permeance W (kg/s m^2 Pa) is calculated by

$$W = G/(A \cdot \Delta p_v) \quad (4.7)$$

where Δp_v (Pa) is the water vapor pressure difference across the specimen:

$$\Delta p_v - p_{sat}(94\%) - p_{sat}(50\%) \quad (4.8)$$

where p_{sat} (94%) is the vapor pressure inside the test cup (Pa), and p_{sat} (50%) is the vapor pressure inside the climatic chamber (Pa);

$$p_{sat} = \varphi \cdot 610,5 \cdot e^{((17,269 \cdot \vartheta)/(273,3+\vartheta))} \quad (4.9)$$

φ is the relative humidity (%); ϑ is the temperature (°C).

The water vapor resistance Z ((s·m^2·Pa)/kg) is the inverse of the water vapor permeance:

$$Z = 1/W \quad (4.10)$$

The water vapor permeability δ (kg/(m·s·Pa)) is given by

$$\delta = W \cdot d \quad (4.11)$$

where d is the mean thickness of the specimen (m).

The water vapor resistance factor μ is calculated through

$$\mu = \delta_a/\delta \quad (4.12)$$

where δ_a is the water vapor permeability of air (kg/(m·s·Pa)) given by

$$\delta_a = ((0.086\, p_0)/(R_v \cdot T \cdot p)) \cdot (T/273)^{1.81} \quad (4.13)$$

where R_v is the gas constant for water vapor, equal to 462 (Nm/kg K), T is the thermodynamic temperature (K), p_0 is the standard barometric pressure of 1013.25 (hPa), and p is the mean barometric pressure over the test (hPa).

The water vapor diffusion-equivalent air layer thickness, s_d (m), is calculated as

$$s_d = \mu \cdot d \quad (4.14)$$

4.2.6 Spectrophotometric Tests

Spectrophotometry is the quantitative measurement of the reflection or transmission properties of a material as a function of wavelength. Therefore, it is possible to identify the transmittance and reflectance properties of a material in visible and solar ranges and the emissivity properties in the medium and far infrared regions.

The instruments (Fig. 4.11) used are three different spectrophotometers that analyze the samples in different incident wavelengths:

- for visible and solar ranges Jasco V-670 UV–VIS–NIR spectrophotometer.
- for medium and far infrared regions Perkin–Elmer Spectrum GX FTIR System.

4.2.7 Scanning Electron Microscopy

The objective of the test is to make morphological analyses with higher resolution than the optical microscope.

The test was carried out with the aid of a SEM PHILIPS XL20 scanning electron microscope (Fig. 4.12). Higher magnified analyses useful for materials containing nanoparticles (to identify nanoparticles average diameter and length) were conducted with Zeiss Supra 40.

The samples were cut into small pieces and attached to an aluminum SEM stub. The specimens were taken with dimensions of 10 × 10 mm^2 to allow gluing on a 10 mm diameter stub. The specimens must be conductive, so they were coated with gold by a plasma sputter to prevent charge build-up by the electron to be adsorbed by the specimens. Subsequently, the samples

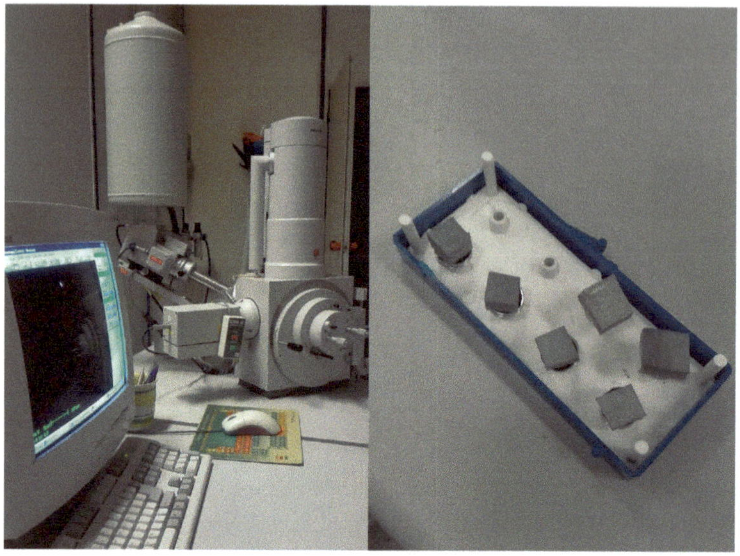

FIG. 4.12 Equipment and samples for SEM test.

were moved into the chamber and micrographs were taken at an accelerating voltage of 20 kV using Philips XL20, whereas of 7 kV for Zeiss Supra 40.

For polyurethane foams, the direction of the foaming process determined different morphologies. Therefore, the morphological analyses were carried out both on the direction of the foam rise and perpendicularly to such direction to evaluate the differences.

The test procedure involves the use of a focused beam of high-energy electrons to generate a variety of signals at the surface of solid specimens. The signals that derive from electron–sample interactions reveal information about the sample including external morphology (texture), crystalline structure, and orientation of materials making up the sample. The SEM is also able to identify chemical compositions by acquiring spot analyses using EDS (Energy-Dispersive X-Ray Spectroscopy).

4.2.8 Aging in Hot-Humid Environment

The aim of the environmental aging tests carried out in this study was to promote the condensation of humidity on the surface of the test specimen by exposing it to an environment with constant air temperature of 40°C and a constant relative humidity of 100%. The standard adopted is ISO 6270-2 (CH method); a climatic chamber "Angelantoni" CST- 130 S model was used for this purpose.

The test equipment provides the use of a vapor-tight climatic chamber installed inside a laboratory room with 23°C temperature and maximum relative humidity of 75%. The adopted chamber is transparent to natural light. A water-filled receptacle (with water depth of 10 mm) is installed in the bottom floor of the chamber.

The test procedure provides the insertion of the samples within the chamber at 40°C temperature, with at an angle of 60 degrees with respect to the horizontal plane to permit drainage of condensed water. Precautions were taken to avoid that the condensation water drip onto the test specimens, neither from the walls or ceiling of the climatic chamber nor from other specimens. The distance from the walls was greater than 100 mm, the distance from the surface of the water was 250 mm and the spacing between specimens was not less than 20 mm.

The test duration was determined for each type of material by identifying on other literature studies the performance decrease trends. For example, for composite materials, the aging lasted for 3 months, while for smart coatings for glazing only 3 weeks. Subsequently, specimens were removed from the climatic chamber and immediately tested (i.e., to measure the mechanical performance after aging), without any further conditioning.

For the tests on glasses (Fig. 4.13), the dimensions of the samples were 10 cm × 15 cm. The samples were obtained by cutting commercial insulated glazing units and were subjected to the specific conditions. For the first 4 days, after 24-hour cycles, the samples were observed using a Wild Makroskop M420 stereo microscope and an optical zoom 12.5X. A raster graphics

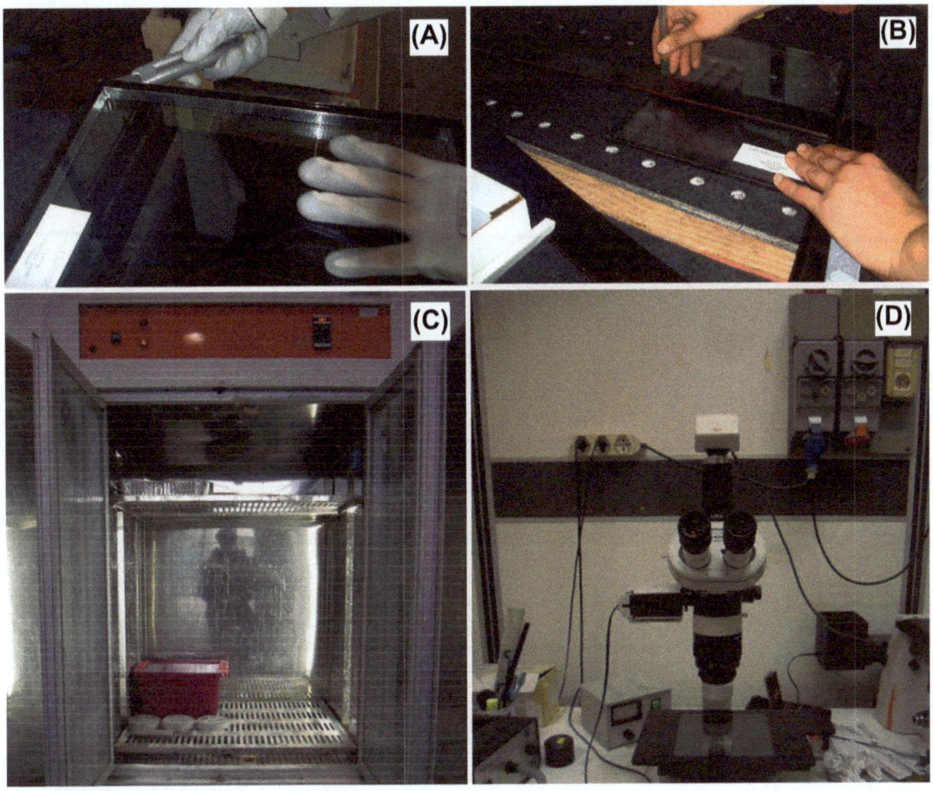

FIG. 4.13 Operations of a low emissivity IGU disassembly **(A, B)**; image of the climatic chamber **(C)**; microscope observations **(D)**.

FIG. 4.14 Salt spray machine of Angelantoni Industrie SPA.

software (Autocad) was employed for the pixel count to evaluate the degradation and formation of white dots representing the oxidation, with consequent detachment of the coating. At the 21st day, the dimensions of white dots were measured, with the purpose of comparing the degradation levels of the glass coatings.

For pultruded profiles with adhesive joints, the Cycle D2 of EN ISO 9142:2003 was instead selected. Each cycle consisted of three exposure periods: first 16 h at the constant temperature of 40°C and relative humidity of 100%, then 4 h at the constant temperature of 40°C and finally 4 h at the constant temperature of 70°C and relative humidity of 50%. This aging cycle was repeated for 2 weeks, according to the standard.

4.2.9 Aging in Saline Environment

This test consists in subjecting the sample to a wet saline atmosphere at a constant temperature. For coated glass, the adopted standard is EN 1096-2:2012.

The test equipment provides the use of a saline solution in a salt spray machine (Fig. 4.14) able to maintain a 40°C constant temperature. The design of the test was to prevent the sprayed solution droplets formed on the surfaces of the chamber from falling onto the specimens. The test chamber used for the tests

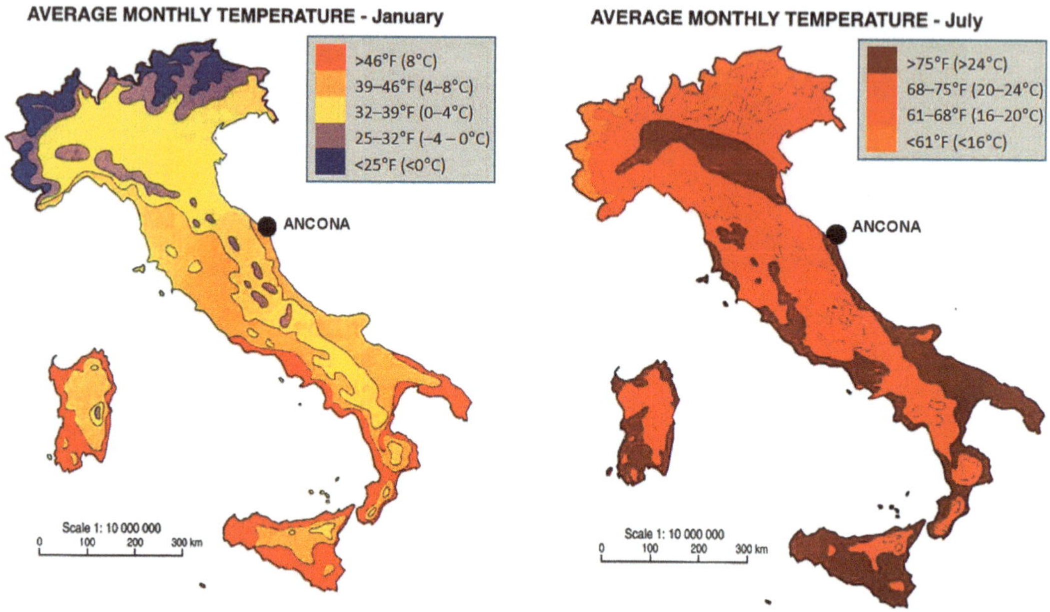

FIG. 4.15 Location and climate of Ancona, the city where the experimentations were carried out.

is a DCTC 600 P salt spray machine of Angelantoni Industrie SPA. The test cabin was operated for a minimum period of 24 hours before the specimens were placed inside. The solution used is water with 3% NaCl.

The coated glass specimens were obtained by cutting samples of 10 cm × 15 cm dimensions from the initial IGUs. The samples were subjected to the specific conditions described earlier for 3 weeks. After the exposure the samples were observed via a stereo microscope allowing to evaluate the degradation and formation of white dots that represent oxidation, resulting in detachment of the coating.

FIG. 4.16 External view of the surveyed university building.

4.3 ON-SITE TESTS

The on-site tests were carried out in real buildings in Ancona, Central Italy. The climate is warm and temperate. There is significant rainfall throughout the year, even in the driest month. According to Köppen and Geiger, the climate classification is Cfa, namely temperate climate with hot summer. The temperature in Ancona (Fig. 4.15) shows average winter values down to 4–8°C (6.6 °C in January according to UNI 10349-1) and warm conditions in summer (average of 25.7 °C in July).

4.3.1 Office Building

The experimental studies regarding office usage typology were carried out in a university building that includes both offices and classrooms (Fig. 4.16).

For the experimental campaign on occupants' behavior, the rooms were adjacent and northern-oriented and each one hosted three about 30-year-old people, equally distributed between male and female (Fig. 4.17). The net floor area was 20 m^2, and the heated volume was 60 m^3 for each room. The glazed surface of each room was 6 m^2, including 3 m^2 of openable casings.

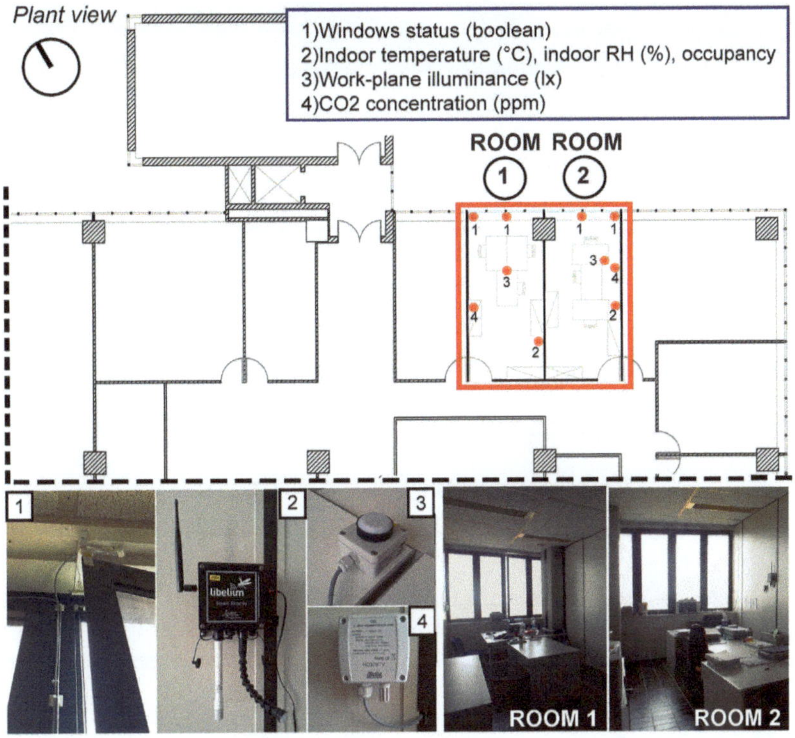

FIG. 4.17 Building plant with probes position; (1, 2, 3, 4) details of probes; images of the rooms 1 and 2.

The exposed surface to volume area, S/V ratio, was 0.33 for both rooms. The occupied period is variable along the week and months. However, the offices are typically occupied from Monday to Friday, from 9 a.m. to 6:30 p.m. The external facade is characterized by strip windows and thermal break aluminum frames. The double-glazing is a solar control low-e glass with 4/9/4 mm IGUs. The opaque vertical envelope consists of a 43 cm thick insulated cavity wall.

The monitoring lasted 1 year, from May 2016 to May 2017, allowing acquisitions of data for all the seasons.

A dedicated sensors network based on an IoT Gateway was implemented with two different sensing nodes in each room (Table 4.1): one measuring air temperature, relative humidity, and occupancy; another one measuring CO_2 concentration, indoor illuminance, and windows opening. The outdoor temperature was recorded from a weather station located at about 2 km from the surveyed building. The window openings were recorded through four magnetic contact-sensors for each room, one for each casement through a Boolean output (i.e., 0 = window closed and 1 = window open). The subjects also filled out nameless questionnaires to register subjective sensations on the thermal environment.

For experimental campaign on solar shadings, adjacent south-western facing rooms with analogue dimensions, the same window covered surfaces and exposure were chosen within the same building. For each room, the net floor area was 28 m^2 and the heated volume was about 95 m^3. The glazed surface was 8.6 m^2, subdivided in six openable windows with dimensions 0.75 m × 1.95 m (Fig. 4.18). Different solar shadings have been installed: aluminum horizontal louvers, aluminum sliding perforated panels, traditional aluminum and wooden persiana, and external low-emissivity shade. Moreover, a ventilated double-glazed window was installed.

The monitoring lasted 3 months, from September 17 until September 29, and from October 24 until November 6. The list of the adopted sensor is reported in Table 4.2.

The rooms were also simulated with Energy Plus software and the model was calibrated with experimental data (Fig. 4.19). The heating and cooling set points were fixed, respectively, at 20°C and 26°C. The

TABLE 4.1
Main Features of the Adopted Probes for the survey on occupants' behaviour in offices.

Sensor	Parameter	Range	Accuracy
Thermistor (SHT75)	Air temperature [°C]	0 ÷ 70 °C	±0.4 °C
Capacitive (SHT75)	RH [%]	0 ÷ 100%	±1.8%
NDIR	CO2 [ppm]	0 ÷ 2000 ppm	±50 ppm
Photodiode Si	Light [lux]	0.02 ÷ 20 klx	±3%
PIR	Num. Occupants	12 m	n/a
Magnetic	Num. Windows open	n/a	n/a
Pt100	Air temperature [°C]	−30C+ 100°C	±0.1°C
Globe thermometer	Mean radiant temp. [°C]	−30C+ 120°C	±0.2°C
Omni direct. hot wire	Air velocity [m/s]	0.05–1 m/s 1–5 m/s	±0.02 m/s ±0.1 m/s
Capacity	Relative humidity [%]	5%–98%	±2.5%

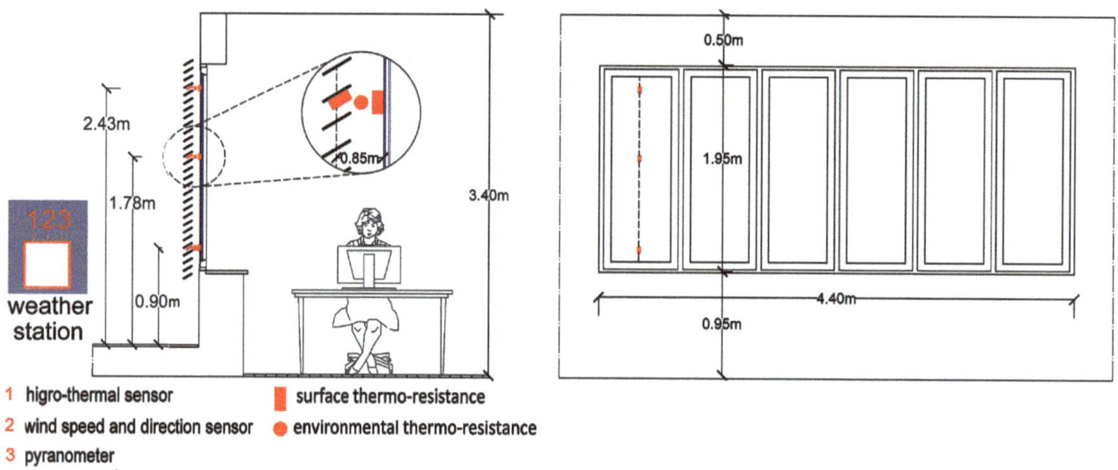

FIG. 4.18 Section and front scheme of the strips windows. (From Stazi et al. (2014).)

ventilation rate was set at 0.3 vol./hour (according to UNI/TS 11300:2008, part 1).

The occupancy was simulated considering 0.1 person/m^2 doing light office work activities (0.9 met).

4.3.2 School Building

The experimentations on the comparison between automatized window use and adaptive free use were carried out in a school building (Fig. 4.20).

Two adjacent classrooms (called "freely-managed" and "motorized") both situated on the ground floor level were selected (Fig. 4.21). The tested rooms present similar features: they have the same external envelope, exposure (north-west side), and occupation density. The classrooms hosted 14/15-year-old students: 17 scholars in the motorized classroom and 16 in the free running one. The net floor area of each classroom was about 55 m^2, and the heated volume was 176 m^3. The total glazed surface of each classroom was 6.8 m^2, subdivided into two windows with lower casements and hopper windows on the top. The exposed surface area to volume S/V ratio was 0.12 for each one. The occupied period is from Monday to Friday and from 8 a.m. to 2 p.m.

The monitoring lasted about 2 months, from 15th to 28th of January 2013 and from 19th of March to the 15th of April 2015.

In both classrooms, the sensors were placed at 1.1 m above the floor according to ISO 7726:2001 for seated persons, in an almost central position. The following data were collected (Fig. 4.21): indoor and outdoor environmental parameters, occupancy patterns, and users' actions on windows. A detailed list of the used sensors is reported in Table 4.3. Research supervisors daily noted on ad hoc spreadsheets real occupancy data (i.e., number of students and teachers), adaptive actions (i.e., opening/closing door and windows), and contingent events.

A preliminary phase regarded the study of adaptive actions of students of the free running (FR) classroom, where the occupants could interact freely with the windows. The students' actions were analyzed to establish the trigger environmental parameters and to verify the Humphreys' adaptive comfort algorithm. Then, it was chosen as control equation to guide an automatized opening/closing system mounted in the windows.

The automatic system, driven by the algorithm, was then installed in motorized (MT) classroom: hopper windows were moved by the installed devices while casement windows use was interdicted to the occupants. Parallel monitoring in both the classrooms (FR and MT) allowed comparing the indoor conditions.

TABLE 4.2
Main Features of the Adopted Probes for the comparison between different solar shadings.

Sensor	Parameter	Accuracy
Thermo resistances	Air temperature (°C)	±0.15°C (at 0°C)
Hygrothermal probe	Air temperature (°C) RH (%)	±0.15°C (at 0°C); ±2% (5%–95%, 23°C).
Hot sphere anemometer	Air velocity (m/s)	±0.03 m/s + 5% m.v./reading
Pyranometers	Light (lux)	0.5% m.v. + 5 W/m^2
Wind direction probe	Wind direction (°)	5°
Wind speed probe	Wind speed (m/s)	2.5% m.v./reading
Data logger	–	3% m.v./reading.

4.4 MOCK-UP

The experimental studies regarding the comparison between techniques in controlled indoor conditions were carried out in a mock-up in Ancona (Figs. 4.22 and 4.23).

The mock-up is a rectangular room of 3.2 m × 5.2 m representative of a full-scale indoor environment.

The test room was realized with a cross-laminated load-bearing timber structure insulated on both external and internal sides. No windows were realized in the mock-up, to avoid uncontrollable heat gains. Realistic indoor gains (occupancy and equipment) were supplied through electric radiators. The entire system of sensors and actuators, as well as the dataflow management, was demanded to a virtual instrument (VI), specifically developed in LabVIEW (programming platform by National Instruments).

The gains simulator was incorporated as a stand-alone subroutine mimicking the internal loads of a residential unit room in summertime. The mock-up was simulated with Ecotect (Fig. 4.24) for the reproduction of sun paths and for the selection of window most critical exposure that resulted in the western side.

To define the amount of total solar transmittance through a westbound glazing surface, the model of

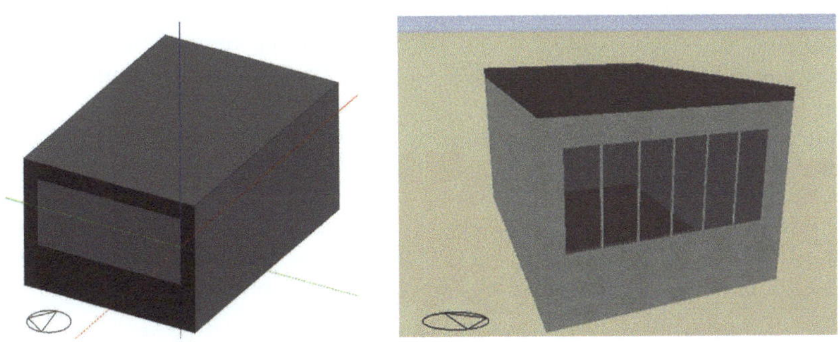

FIG. 4.19 Simulation models of one room by reproducing the real conditions.

FIG. 4.20 Images of the case study (southern side).

the mock-up was reconstructed on Sketch Up, using the Open Studio plug-in and a dynamic thermal simulation was carried out with Energy Plus. The model was calibrated with the monitored outdoor temperatures and solar radiation level.

Under such real boundary conditions, the curves of the total solar transmittance due to a west-oriented window were obtained by parametric simulation (Fig. 4.25). In detail, a generic medium-sized window (1.5 m^2) facing west was introduced in the model. Indeed, this window size is typically of a residential unit in proper natural lighting (glazed area higher than 1/8 fraction of the floor area).

The relative values were converted in solar gains (W/m^2) and adopted as heat to supply. In particular, two levels of heat transmitted through a window have been chosen: a medium level, from the typical solar transmittance obtained for cloudy days (Test 1) and a high level, considering the typical solar transmittance on sunny days (Test 2).

The curves simulate the warming of indoor surfaces due to the greenhouse effect.

4.4.1 Internal Envelope Study

The internal side of the south-facing wall of the mock-up was used to simultaneously compare different internal linings (Fig. 4.26).

Four different linings were compared (Fig. 4.27):

L: traditional plasterboard finishing

M: one layer dry clay (2.8 cm thick totally)

H: two layers dry clay (4.2 cm thick totally)

H_I: 4 cm bricks layer combined with 4 cm insulation layer

The yearly on-field monitoring campaign provided the following measurements:

- Outdoor climatic conditions. A weather station was used to measure the external boundary conditions: global solar radiation, dry-bulb air temperature, relative humidity, wind speed, and direction.

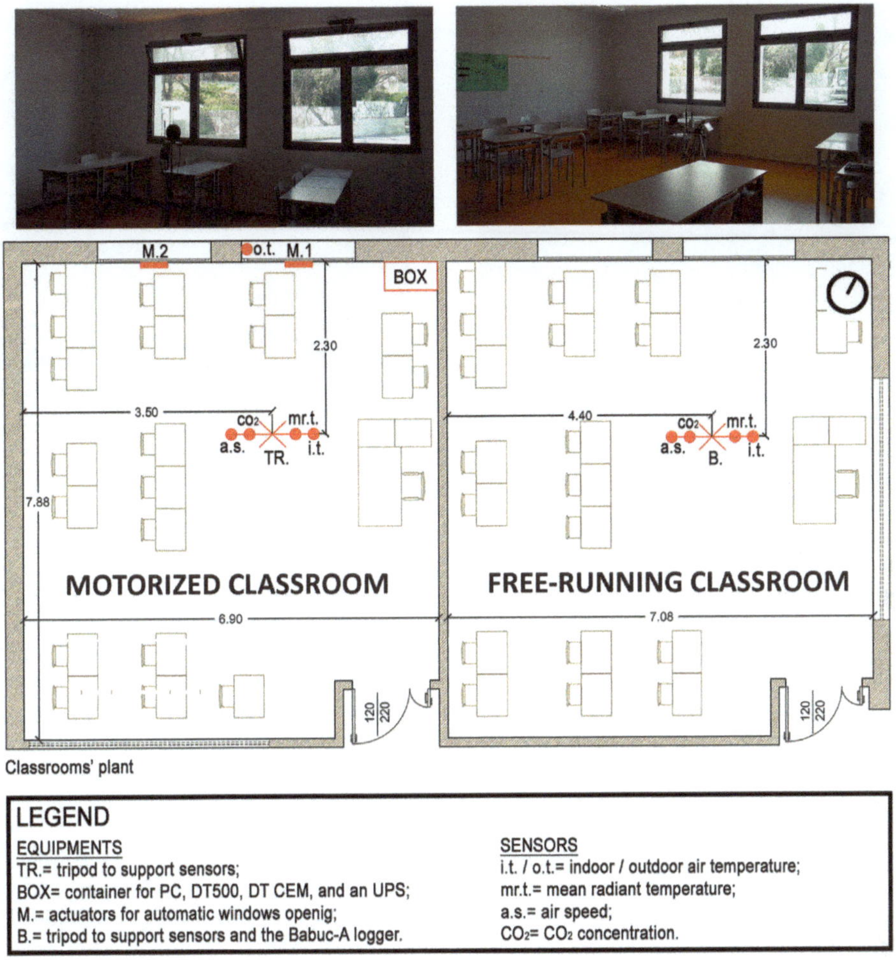

FIG. 4.21 Internal images and plants of the two classrooms with position of the probes. (Modified from Stazi et al. (2017).)

TABLE 4.3
Accuracy of the Probes Installed in the Classrooms.

Sensors	Parameter	Range	Accuracy
Thermoresistances	Indoor air temperature (°C)	−50 ÷ 80°C	0.15°C
Globe therm.	Mean radiant temp. (°C)	−40 ÷ 80°C	0.15°C
Anemometer	Indoor air speed (m/s)	0.01 ÷ 20 m/s	0 ÷ 0.5 m/s
CO_2 sensor	CO_2 concentration (ppm)	0 ÷ 3000 ppm	20 ÷ 70 ppm

FIG. 4.22 Mock-up view from southern side.

FIG. 4.23 Mock-up view form northern side. (From Stazi et al. (2018).)

- Indoor environment. A microclimate station was placed in the midpoint of the room to detect the thermohygrometric variations of the internal air. The sensors measured wet and dry-bulb temperature.
- Envelope performance. A net of heat flux meters and flat thermistors was implemented on the internal side of the tested linings to map the different surface temperatures and fluxes achieved under the same internal conditions.

The internal air stratification was also evaluated (Fig. 4.28).

Extensive results of the experimentations are also reported in (Stazi et al., 2018).

4.4.2 External Envelope Study

The west-facing wall of the mock-up was used to install the ventilated skin prototypes. Three ventilated skins were simultaneously compared (Fig. 4.29).

The first configuration (L) is the lightweight typology without mass. It has an aerated gap enclosed by a lightweight panel and it was used as reference for comparisons (Fig. 4.30). Screwed vertical wooden battens (6 cm width) and an external white plastered oriented strand board (OSB) panel were used.

The second configuration (IM) has an internal mass.

In this solution, the clay blocks are laid adjacent to the insulation material; the gap is then confined through a white plastered OSB panel, as in the first solution (Fig. 4.31). For the realization of this wall, stainless steel supporting brackets were anchored (also providing a thermal cutting plate) at the cross-laminated structure in the bottom of the walls (Fig. 4.31A); then, the massive layer was assembled adjacent to the existing insulation layer, using also cavity ties distributed for the entire surfaces (Fig. 4.31B and C). Finally, an external plastered OSB was used by leaving an interspace of 6 cm with respect to the massive layer.

The third configuration (EM) has an external white plastered mass.

Stainless steel supporting brackets (as those used in IM wall) were anchored at the cross-laminated structure in the bottom of the walls (Fig. 4.32A); then, the massive layer was assembled leaving an interspace (6 cm width) with the existing insulation layer. Cavity ties (longer than those of IM wall) were adopted for the entire surface to fix the wall; the bricks were then plastered.

Obviously, the stainless steel supporting brackets of EM solution were longer than those of IM (Fig. 4.33).

To reach the same optical properties for all the walls the same external white shaving was used. Honeycombed nets were placed to protect all the cavities. Lateral insulation boards were used to seal and insulate each wall.

The yearly extensive monitoring activity was carried out to examine the thermal behavior of the proposed ventilated skin technologies. The probes were installed either in the internal surfaces of the cavities or in the external surface of each cladding.

The following sensors were used (Fig. 4.34):

- Outdoor temperature, relative humidity, global solar radiation, speed, and direction of the wind. The data were collected using a weather station;
- Surface temperatures of the outermost and innermost layers measured at the bottom (60 cm), at the middle (115 cm), and at the top (168 cm) of each façade. The data were collected adopting thermoresistance sensors (accuracy $\pm$ 0.05°C);

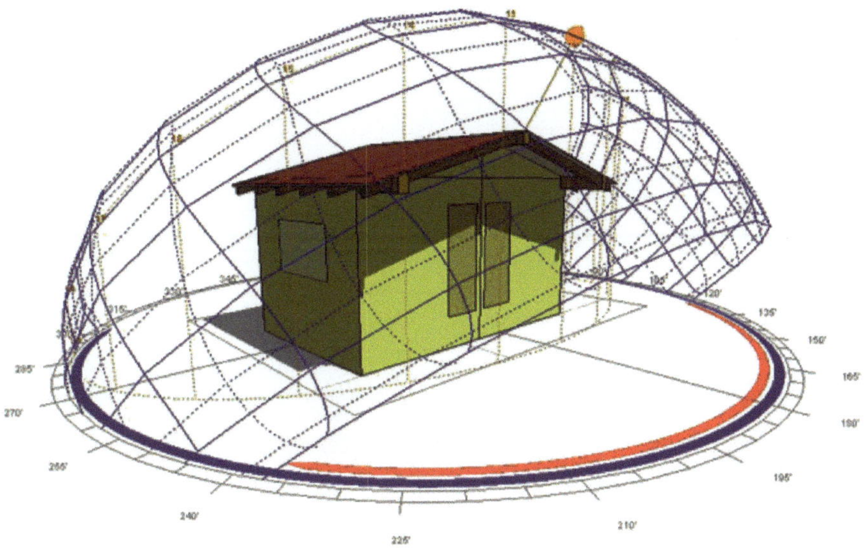

FIG. 4.24 Simulation of the model with Ecotect.

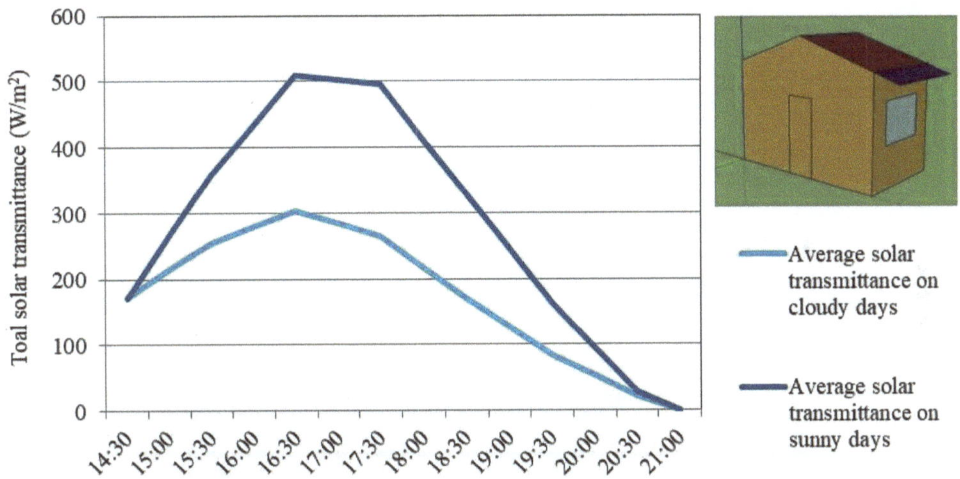

FIG. 4.25 Average simulated total solar transmittance passing through the west window of the mock-up on summer season (Energy Plus software). (From Stazi et al. (2018).)

- Incoming and outgoing heat fluxes measured at mid-height (115 cm) in the innermost side of the air chamber. Data were collected through heat flux plates (precision ± 3%);
- Air velocity and air temperature inside the ventilation chamber recorded at mid-height (115 cm). The data were collected using hot-sphere anemometers (tolerance of ±0.03 m/s).

The data acquisition involved gathering analog signals from measurement sources and digitizing them, adopting NI-DAQ hardware for storage. LabVIEW software was used and implemented for real-time data presentation. The sample rate of the data acquisition system was set every 5 minutes.

4.5 NUMERICAL CALCULATIONS: THERMAL TRANSMITTANCE OF A WINDOW

There are two different ways to calculate the thermal transmittance of a window. The two approaches

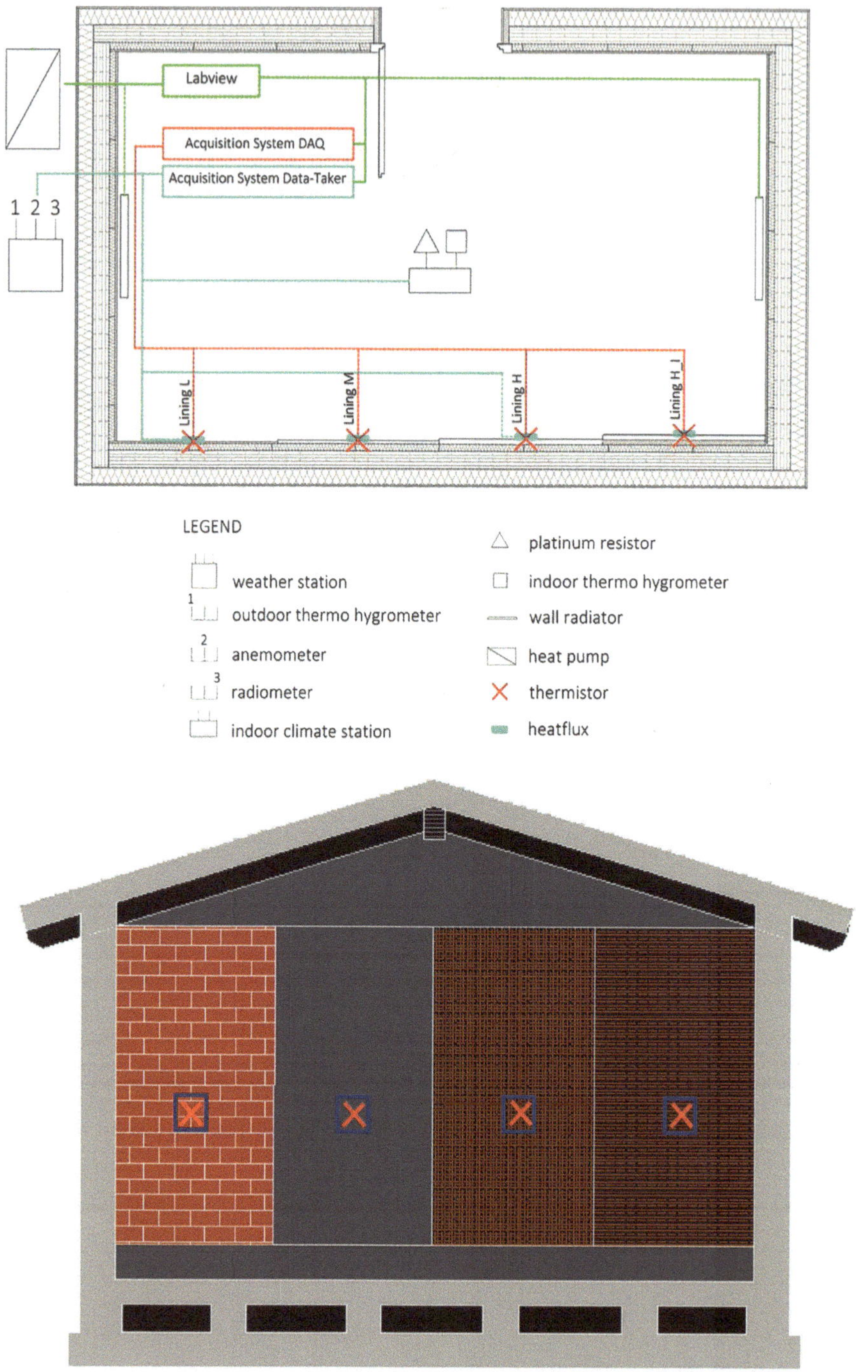

FIG. 4.26 Plant and internal front of the wall with monitoring probes.

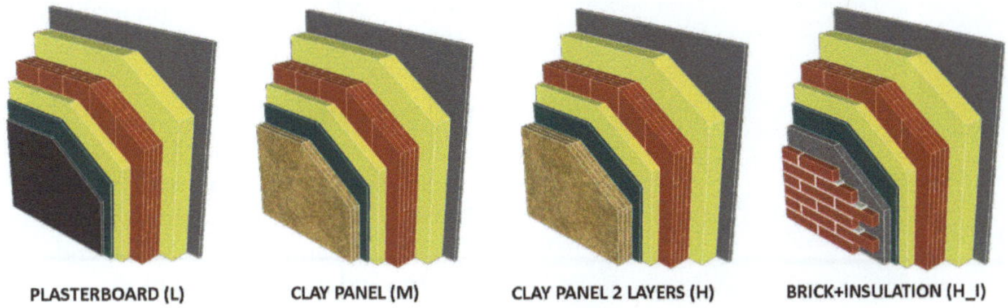

FIG. 4.27 Tridimensional views of the compared linings.

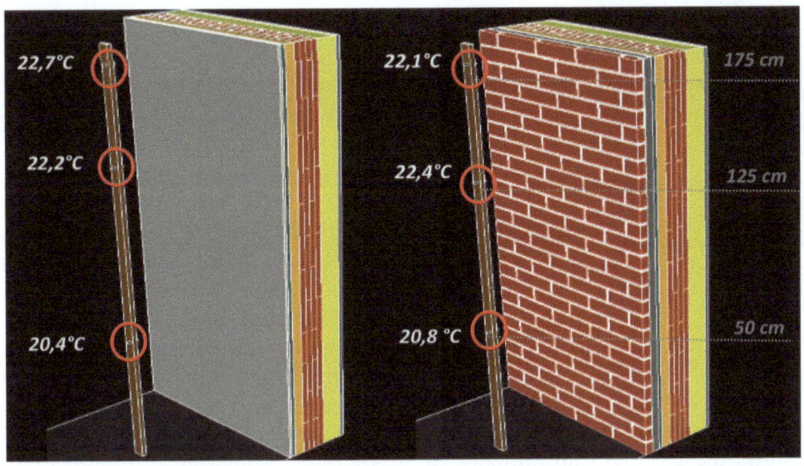

FIG. 4.28 Position of the probes to monitor the thermal stratification of the air adjacent to the linings.

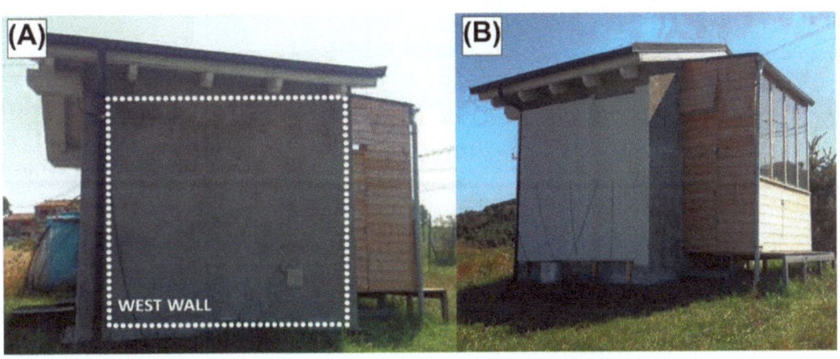

FIG. 4.29 West view of the mock-up before **(A)** and after **(B)** the vented facades realization.

(referred as ASHRAE procedure and ISO procedure) are both described in ISO 15099 and differ in the way they treat the effect of the glazing spacer on heat transfer through the frame and the glazing unit near the frame (Blanusa et al., 2007).

ASHRAE procedure

It provides the thermal transmittance calculation as the area-weighted average of the thermal transmittance of the frame (*f*), edge of glass (*eg*), and center of glass (*cog*). The "*eg*" is the perimeter area of

FIG. 4.30 Realization phases of lining L with wooden battens and OSB panel (before plastering).

the IG unit containing the spacer and sealants, while the *cog* comprises the vision area of the glazing (Fig. 4.35).

The equation to find the thermal transmittance of the window U_w is then the following:

$$U_{w,ASHRAE} = \frac{U_{cog}A_{cog} + \sum U_f A_f + \sum U_{eg}A_{eg}}{A_t} \tag{4.15}$$

where U_{cog} is the center of glass U-value; A_{cog}, A_f, A_{eg}, A_t are, respectively, the center of glass area, the projected frame area, the edge of glass area, and the total window area. U_f and U_{eg} are obtained with the following formula:

$$U_f = \frac{\Phi_{fr}}{l_f \Delta T} \tag{4.16}$$

$$U_{eg} = \frac{\Phi_{eg}}{l_{eg} \Delta T} \tag{4.17}$$

where l_f is the projected length of the frame area; l_{eg} is the length of the edge of glass area equal to 63.5 mm; Φ_{fr} and Φ_{eg} are the heat flow rates through frames and edge of glass area including the effect of glass and spacer; $\Delta T = T_{in} - T_{out}$, respectively, the internal and external temperatures.

ISO procedure

In this procedure, the frame is simulated with an insulation panel instead of a glazing to find the frame U-value. In particular, it allows calculating a linear thermal transmittance that depends on the adopted spacing and glass configuration.

The overall thermal transmittance of the window U_w can be obtained through the following formula:

$$U_{w,\ ISO} = \frac{\sum(U_g A_g) + \sum(U_f A_f) + \sum(l_\Psi \Psi_g)}{A_t} \tag{4.18}$$

where U_g is the thermal transmittance value in the center of the glazing unit (it coincides with U_{cog}); U_f is the thermal transmittance of the frame; A_g and A_f are projected vision and frame areas; A_t is the total window area; l_Ψ is the length of the vision area perimeter (often called sightline); Ψ_g is the linear thermal transmittance characterizing the additional heat transfer due to the spacer region of the IGU.

The frame thermal transmittance U_f can be determined by simulating the frame with the insertion of an insulation panel (instead of the real glazing and spacer), performing a two-dimensional heat transfer calculation (section of length l_p) and through the following equation:

$$U_f = \frac{(q/\Delta T) - U_p l_p}{l_f} \tag{4.19}$$

where $\Delta T = T_{in} - T_{out}$; U_p is the thermal transmittance of the panel foam insert; l_p is the internal side exposed length of foam insert; l_f is the projected length of the frame section; $q/\Delta T$ is the thermal coupling coefficient determined from the frame-panel insert system.

The linear thermal transmittance Ψ_g is determined as follows:

$$\Psi_g = \frac{q_{total}}{\Delta T} - U_f l_f - U_g l_g \tag{4.20}$$

where $q_{total}/\Delta T$ is the thermal coupling coefficient determined from the actual fenestration system (frame, glazing, and spacer); l_g is the projected length of the glass section.

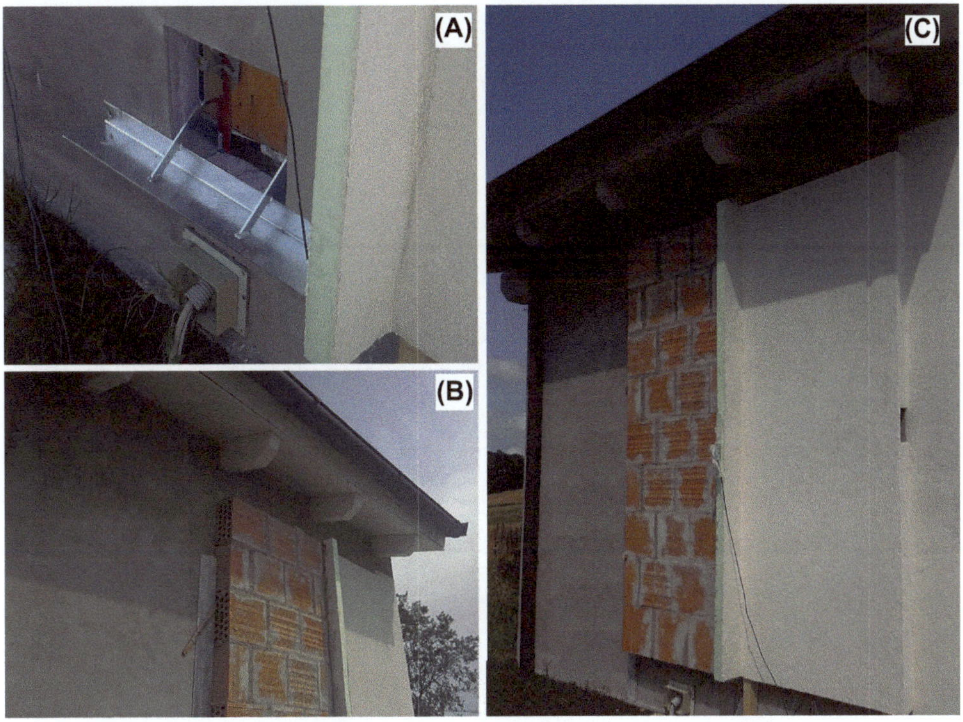

FIG. 4.31 Realization of wall IM: **(A)** stainless steel brackets; **(B)** realization of massive layer with perforated bricks; **(C)** massive layer complete, with fixing ties.

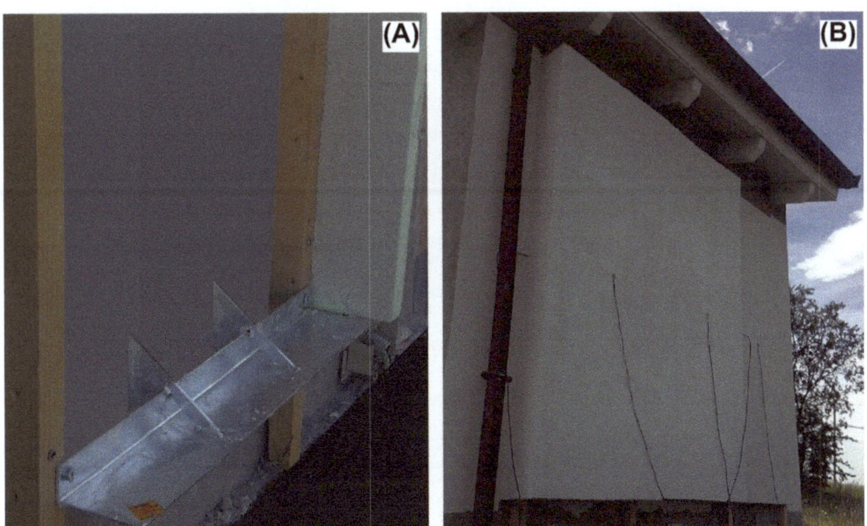

FIG. 4.32 Realization of wall EM: **(A)** stainless steel brackets and **(B)** completed wall.

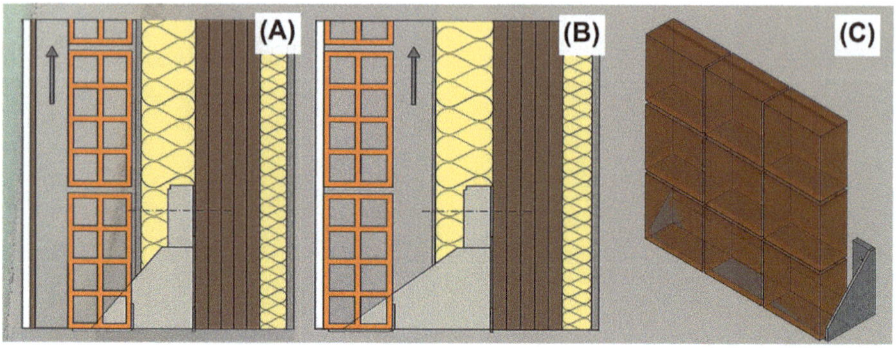

FIG. 4.33 Details of the different stainless steel supports used for IM **(A)** and EM (**B** and **C**).

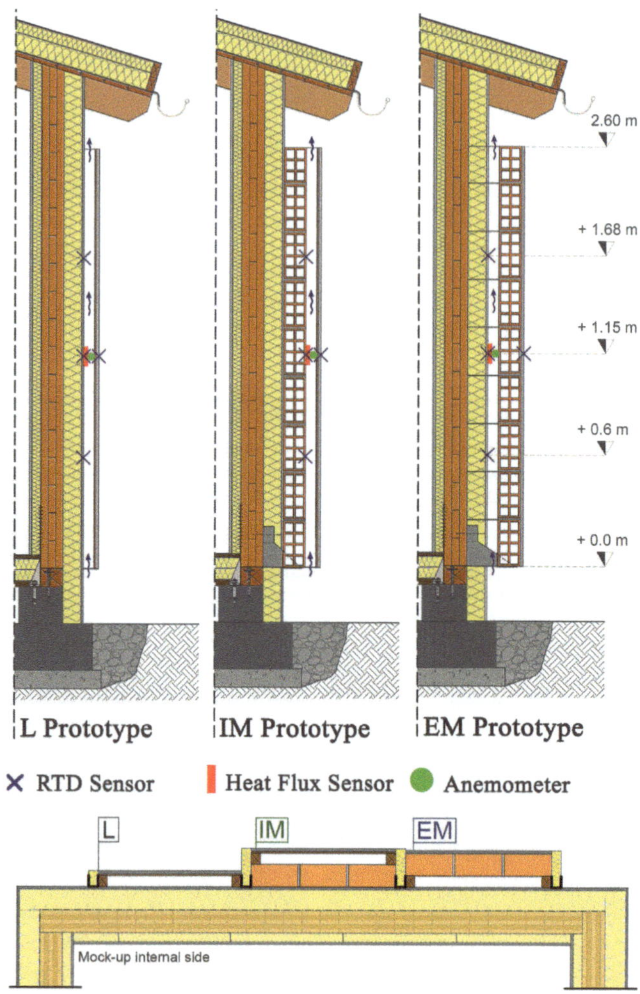

FIG. 4.34 Monitoring plan of the three ventilated prototypes, section, and plant.

4.6 NUMERICAL CALCULATIONS: OPTICAL PROPERTIES OF THE GLAZING

The optical properties of the glazing were calculated according to EN 410:2011.

Solar light factors

Visible light transmittance τ_v (%)

It is the amount of incident light in the visible portion of the spectrum that is transmitted by the glass.

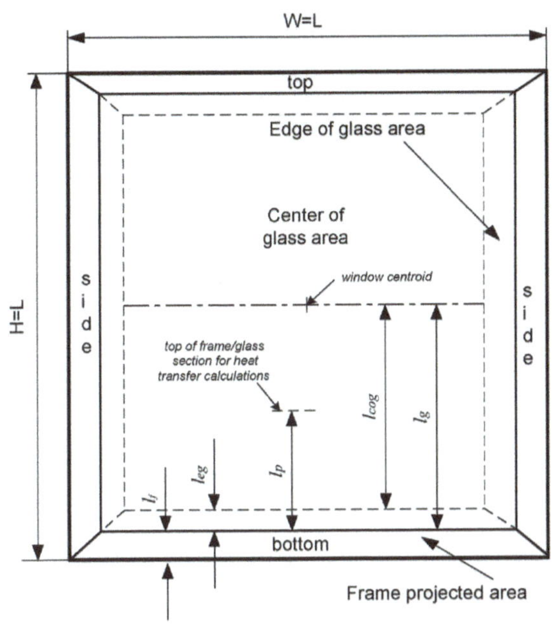

FIG. 4.35 Scheme of window with parameters. (From Blanusa et al. (2007).)

It can be calculated using the following:

$$\tau_v = \frac{\sum_{\lambda=380\,nm}^{780\,nm} D_\lambda \tau(\lambda)\, V(\lambda) \Delta\lambda}{\sum_{\lambda=380\,nm}^{780\,nm} D_\lambda V(\lambda) \Delta\lambda} \tag{4.21}$$

where $\tau(\lambda)$ is the spectral transmittance (%) of the coated glazing experimentally determined; D_λ is the relative spectral distribution of the illuminant D65 (K); $V(\lambda)$ is the spectral luminous efficiency for photopic vision (lm/W) defining the standard observer for photometry; $\Delta\lambda$ is the wavelength interval [nm]. The values for $D_\lambda\ V(\lambda)\ \Delta\lambda$ are obtained from Table 1 of the standard.

It is influenced by the glazing type, the number of panes, and the presence of coatings. Visible transmittance of glazing ranges from above 0.9, for uncoated water-white clear glass to less than 0.10 for highly reflective coatings on tinted glass. This means that 90% and 10%, respectively, of visible light are transmitted. The higher is the τ_v-value the higher is the natural indoor illumination with a consequent reduction on electric lighting consumptions.

Visible light reflectance ρ_v (%)

It is the fraction of the incident light that is reflected by the glass.

It can be calculated using the following formula:

$$\rho_v = \frac{\sum_{\lambda=380\,nm}^{780\,nm} D_\lambda \rho(\lambda) V(\lambda) \Delta\lambda}{\sum_{\lambda=380\,nm}^{780\,nm} D_\lambda V(\lambda) \Delta\lambda} \tag{4.22}$$

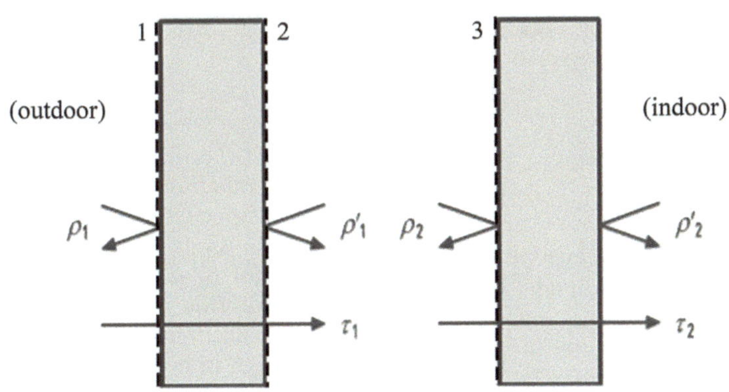

FIG. 4.36 Transmittance and reflectance in a double-glazing insulating glass unit. (From Stazi et al. (2016).)

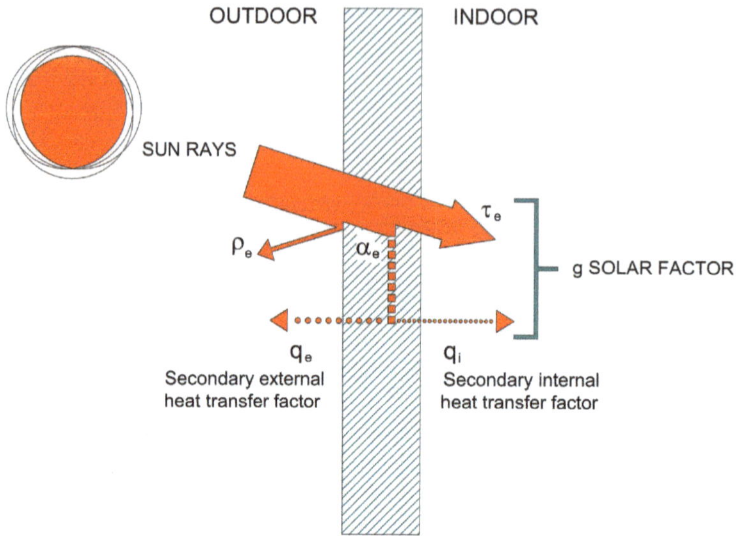

FIG. 4.37 Scheme of the Total Solar Energy Transmittance g.

where $\rho(\lambda)$ is the spectral reflectance (%) of the coated glass experimentally determined; D_λ is the relative spectral distribution of the illuminant D65 (K); $V(\lambda)$ is the spectral luminous efficiency for photopic vision [lm/W] defining the standard observer for photometry; $\Delta\lambda$ is the wavelength interval [nm]. The values for D_λ $V(\lambda)$ $\Delta\lambda$ are obtained from Table 1 of the standard UNI EN 410. This parameter has the opposite behavior of the light transmittance τ_v, so it should be beneficially reduced.

Solar energy factors

Solar transmittance τ_e (%)

It is the fraction of solar radiation transmitted directly through the glass.

It can be calculated using the following formula (according to UNI EN 410:2011):

$$\tau_e = \frac{\sum_{\lambda=300\,nm}^{2500\,nm} S_\lambda \tau(\lambda)\Delta\lambda}{\sum_{\lambda=300\,nm}^{2500\,nm} S_\lambda \Delta\lambda} \tag{4.23}$$

where $\tau(\lambda)$ is the spectral transmittance of the coated glass experimentally determined (%); S_λ is the relative spectral distribution of the solar radiation [W/m^2]; $\Delta\lambda$ is the wavelength interval [nm]. The values for S_λ $\Delta\lambda$ can be obtained from Table 2 of the standard UNI EN 410.

This parameter affects the calculation of the total solar energy transmittance (*g*-value); therefore, in hot climates, the glass should have low τ_e to keep the solar heat gains as low as possible.

Note that t_e parameter presents a mediated formula between 300 and 2500 nm, including both the visible and the infrared ranges.

Solar reflectance ρ_e (%)

It is the fraction of the incident solar radiation reflected by the glass.

It can be calculated using the following formula (according to UNI EN 410:2011):

$$\rho_e = \frac{\sum_{\lambda=300\,nm}^{2500\,nm} S_\lambda \rho(\lambda)\Delta\lambda}{\sum_{\lambda=300\,nm}^{2500\,nm} S_\lambda \Delta\lambda} \tag{4.24}$$

where ρ (λ) is the spectral reflectance (%) of the coated glass experimentally determined; S_λ is the relative spectral distribution of the solar radiation (W/m^2); $\Delta\lambda$ is the wavelength interval (nm). The values for S_λ $\Delta\lambda$ are obtained from Table 2 of the standard. This parameter has the opposite behavior of the solar transmittance τ_e, so it should be beneficially high in hot climates.

Normal emissivity ε_n (–)

It is the ratio, in a direction normal to the surface, of the emissive power of the surface of the glass to the emissive power of a black body (a perfect emitter). It can be calculated using the following formula (by adopting the EN 12898):

$$\varepsilon_n = 1 - R_n \tag{4.25}$$

where R_n is the infrared total normal reflectance (with a reduced spectrum of 5000 ÷ 25,000), deduced from experimental tests in accordance with formulas (1) and (2) of the standard. The corrected emissivity can be calculated according to Table A.2 of the standard EN 12898. The lower the ε_n, the higher will be the reflecting ability of the pane. This parameter should be beneficially reduced to reflect the room longwave infrared radiations back to the interior in winter and to avoid the undesired absorption of the external heat gains into the glass pane in summer.

IGU spectral transmittance and reflectance

For each glazed insulating units (Fig. 4.36), composed by two single glasses, an internal air gap and the film in the appropriate position, the main optical and energy parameters are in the visible range, $T_v(\lambda)$, $R_v(\lambda)$ and in the solar range, $T_e(\lambda)$, $R_e(\lambda)$.

The values were obtained, respectively, from the spectral transmittances and reflectance of the individual glasses using the following formulas (according to UNI EN 410:2011):

$$T_v(\lambda) = [\tau_{v1}(\lambda)\tau_{v2}(\lambda)]/[1 - \rho'_{v1}(\lambda)\rho_{v2}(\lambda)] \tag{4.26}$$

$$T_e(\lambda) = [\tau_{e1}(\lambda)\tau_{e2}(\lambda)]/[1 - \rho'_{e1}(\lambda)\rho_{e2}(\lambda)] \tag{4.27}$$

$$R_v(\lambda) = \rho_{v1}(\lambda) + \frac{\tau^2{}_{v1}(\lambda)\rho_{v2}(\lambda)}{1 - \rho'_{v1}(\lambda)\rho_{v2}(\lambda)} \tag{4.28}$$

$$R_e(\lambda) = \rho_{e1}(\lambda) + \frac{\tau^2{}_{e1}(\lambda)\rho_{e2}(\lambda)}{1 - \rho'_{e1}(\lambda)\rho_{e2}(\lambda)} \tag{4.29}$$

where $\tau_{v1}(\lambda)$ and $\tau_{e1}(\lambda)$ are, respectively, the light and solar transmittance of the first (outer) pane (%); $\tau_{v2}(\lambda)$ and $\tau_{e2}(\lambda)$ are, respectively, the light and solar transmittance of the second pane (%); $\rho_{v1}(\lambda)$ and $\rho_{e1}(\lambda)$ are, respectively, the light and solar reflectance of the first (outer) pane measured in the direction of the incident radiation (%); $\rho'_{v1}(\lambda)$ and $\rho'_{e1}(\lambda)$ are, respectively, the light and solar reflectance of the first (outer) pane measured in the direction opposite to the incident radiation (%).

Total solar energy transmittance *g*

It is also known as solar factor. It is the solar transmittance plus the absorbed part of the radiation reemitted toward the inside of the building, and thus, it represents the fraction of the incident solar radiation that is totally transmitted by the glass (Fig. 4.37). Despite having minor differences in modeling standards compared to the solar heat gain coefficient (SHGC, mainly used in the United States), the two values are effectively the same. A *g*-value of 1.0 represents full transmittance of all solar radiation while 0.0 represents a window with no solar energy transmittance. Most *g*-values will range between 0.2 and 0.7, with solar control glazing having a *g*-value of less than 0.5.

It can be calculated using the following formula:

$$g = T_e(\lambda) + q_i \tag{4.30}$$

where $T_e(\lambda)$ is the solar transmittance and q_i is the secondary heat internal transfer factor calculated in accordance to paragraph 5.4.6 of the standard UNI EN 410.

This parameter denotes the share of the incoming solar energy, which is converted into heat inside the indoor space.

The lower the *g*-value the less is the transmitted solar heat and the greater is the ability of the IGU at blocking heat gains from the sun.

A product with a high *g* rating is more effective at collecting solar heat during the winter, while a product with a low *g* is more effective at reducing cooling loads during the summer by blocking the heat gain from the sun. The optimal *g*-value depends on the specific climate, orientation, and external shading.

REFERENCES

Blanusa, P., Goss, W. P., Roth, H., Weitzmannn, P., Jensen, C. F., Svendsen, S., & Elmahdy, H. (2007). Comparison between ASHRAE and ISO thermal transmittance calculation methods. *Energy and Buildings, 39*(3), 374–384. https://doi.org/10.1016/j.enbuild.2006.09.007.

Cattarin, G., Causone, F., Kindinis, A., & Pagliano, L. (2016). Outdoor test cells for building envelope experimental characterisation - a literature review. *Renewable and Sustainable Energy Reviews*. https://doi.org/10.1016/j.rser.2015.10.012.

Goia, F., Schlemminger, C., & Gustavsen, A. (2017). The ZEB Test Cell Laboratory. A facility for characterization of building envelope systems under real outdoor conditions. *Energy Procedia*. https://doi.org/10.1016/j.egypro.2017.09.718.

Heydarian, A., Carneiro, J. P., Gerber, D., Becerik-Gerber, B., Hayes, T., & Wood, W. (2015). Immersive virtual environments versus physical built environments: A benchmarking study for building design and user-built environment explorations. *Automation in Construction*. https://doi.org/10.1016/j.autcon.2015.03.020.

Stazi, F., Marinelli, S., Di Perna, C., & Munafò, P. (2014). Comparison on solar shadings: Monitoring of the thermo-physical behaviour, assessment of the energy saving, thermal comfort, natural lighting and environmental impact. *Solar Energy, 105*. https://doi.org/10.1016/j.solener.2014.04.005.

Stazi, F., Giampaoli, M., Tittarelli, F., Di Perna, C., & Munafò, P. (2016). Durability of different glass coatings in humid and saline environments, ageing impact on heat-light transmission and thermal comfort. *Building and Environment, 105*. https://doi.org/10.1016/j.buildenv.2016.05.029.

Stazi, F., Naspi, F., Ulpiani, G., & Di Perna, C. (2017). Indoor air quality and thermal comfort optimization in classrooms developing an automatic system for windows opening and closing. *Energy and Buildings, 139*. https://doi.org/10.1016/j.enbuild.2017.01.017.

Stazi, F., Ulpiani, G., Pergolini, M., & Di Perna, C. (2018). The role of areal heat capacity and decrement factor in case of hyper insulated buildings: An experimental study. *Energy and Buildings, 176*, 310–324. Elsevier B.V. https://doi.org/10.1016/j.enbuild.2018.07.034.

Index

Note: Page numbers followed by "f" indicate figures and "t" indicate tables.

FSC
www.fsc.org
MIX
Papier | Fördert gute Waldnutzung
FSC® C083411

Kolibri 360 GmbH
Ferdinand-Jühlke-Straße 7
99095 Erfurt, Deutschland
produktsicherheit@kolibri360.de